T0290964

PERSPECTIVES IN OPTICS RESEARCH

LASERS AND ELECTRO-OPTICS RESEARCH AND TECHNOLOGY

Additional books in this series can be found on Nova's website under the Series tab.

Additional E-books in this series can be found on Nova's website under the E-book tab.

LASERS AND ELECTRO-OPTICS RESEARCH AND TECHNOLOGY

PERSPECTIVES IN OPTICS RESEARCH

JEFFREY M. RINGER
EDITOR

Nova Science Publishers, Inc.
New York

LIBRARY OF CONGRESS CATALOGING-IN-PUBLICATION DATA

Perspectives in optics research / [edited by] Jeffrey M. Ringer.
    p. cm. -- (Lasers and electro-optics research and technology)
  Includes bibliographical references and index.
  ISBN 978-1-61122-934-9 (hardcover : alk. paper)
  1. Optics. I. Ringer, Jeffrey M.
  QC355.3.P47 2011
  621.36--dc22
                        2010044752

Published by Nova Science Publishers, Inc. † New York

CONTENTS

PREFACE

This book presents and reviews research in the field of optics, including multi-wavelength semiconductor laser diodes; in-line hologram reconstruction; photoacoustic spectroscopy of NO2; chemical oxygen iodine laser; optical superlattices and chalcones.

Chapter 1 – Due to the rapid technical advances of semiconductor laser diodes as a light source, the ultra-high-density optical information storage systems have been accelerated to commercialize in various formats, such as CD (Compact Disk), CD-ROM (Read-Only Memory), CD-R (-Recordable), CD-RW (-ReWriteable), DVD (Digital Versatile/Video Disc), DVD-R, and DVD-RW, for many applications, i.e., audio, music, image, movie, and computer data storage. In this complex situation, the compatibility of diverse storage media discs in a playback/writing system is one of the most important issues for system development. This problem has caused complications in the optical components as well as the non-compact system size. A semiconductor laser is an integral part among many optical components because the wavelength and output power of laser light source is a key factor of the media disc specifications.

In order to realize signal compact optical pick-up module that can respond to various media discs, a multiple wavelength laser light source is indispensable. There have been many attempts to develop multi-wavelength laser diodes. The typical approaches are to use two individually packaged laser diodes or hybrid-integrated two laser chips in the same package. However, those techniques require sophisticated assembly processing resulting in low reliability and high cost.

In this chapter, the authors reviewed the recent progress of multi-wavelength laser diodes and how it can effectively realize dual or multiple wavelength laser diodes in the future. The authors proposed, and successfully demonstrated, new concept device structures for monolithically-integrated two laser diodes with different wavelengths on a GaAs wafer in parallel and vertical direction.

Chapter 2 – The authors analyze the propagation of soliton pulses in an absorbing non-linear three level medium in lambda configuration. One of the two atomic transitions is excited by a strong CW control laser light, whereas the other atomic transition is pumped by a weak variable light field. The authors take into consideration the effects of detunings between the laser sources and the two atomic transitions. Assuming two-photon resonance, the authors present an analytical expression of soliton shape. The soliton propagation velocity is influenced by three parameters: the amplitude of the stationary control field, the maximum amplitude of the soliton and by the detuning. Especially, the authors show that for a given value of the detuning, the soliton can be stopped.

Chapter 3 – Digital holography is a quickly developing technology for high speed imaging. In-line holography has been extensively investigated due to it can effectively utilize the space-bandwidth product of digital recording instruments. However, since the reference wave and the object wave are overlapped during the hologram is recorded; the directly inverse Fresnel transform can not present the original object well. The reconstructed image is blurred by its conjugated image and the zero-order diffraction of the hologram. This shortcoming limits the application of the in-line holography. Therefore, an effective reconstruction algorithm is important for generalizing its application.

In this chapter, some approaches based on the phase retrieval algorithm for in-line hologram reconstruction are reported. Firstly, the Yang-Gu algorithm and the GS algorithm are used to restructure pure absorption and pure phase object from their in-line holograms, respectively. The differences between these two algorithms are analyzed. Then the GS algorithm is extended to reconstruct whole optical field from double or multiple holograms. At last, a new approach for reconstructing object from a hologram series is presented. Experimental results show that all these methods can reconstruct original object well.

Chapter 4 – Photoacoustic spectroscopy is a very powerful tool for quantitative analysis of gaseous traces. Particularly, devices based on excitation by visible laser radiation allow sensitive detection of NO_2, which is a common urban pollutant. In this chapter, different photoacoustic systems are shown: some of them use resonant excitation of the acoustic modes of a cell by modulated laser radiation, other use Fourier analysis of the time response due to irradiation with pulsed lasers. The development of a simple and compact system for pollution control and its application to real measurements is much of interest. At the same time, very different issues get benefit from a vast experimental work concerning the photoacoustic detection of NO_2: the verification of theoretical models for photoacoustic phenomena, studies of surface phenomena (adsorption at the walls) and the determination of unknown absorption coefficients of certain molecules, which undergo multiphoton excitation, based on a precise determination of the setup calibration constant.

Pulsed photoacoustics is applied to the study of the Q-factor of a cylindrical acoustical cavity for determining the main kind of mechanism of the pressure wave's energy loss. A simple signal processing method for analysis in the frequency domain, which allows high resolution spectroscopy with low-cost equipment, is described.

Synchronous detection is used in photoacoustic setups based on the resonant scheme, which use either an amplitude modulated CW green laser or a high repetition frequency-doubled Q-switched Nd:YAG laser. The characteristics of different systems and their detection limit are shown. Further, the acquisition system is simplified through digitizing the pre-amplified signal from the microphone by the sound card of a PC, giving place to low-cost and easy-handling equipment, ideal for field measurements. Some of these systems are applied to the determination of the NO_2 content in car exhausts and the quality of catalytic converters. Moreover, a one-dimensional model for cell design and optimization of S/N ratio is further verified by measurements on NO_2-air mixtures enclosed in a specially built acoustic cavity with a detached Helmholtz resonator.

Chapter 5 – Chemical Oxygen Iodine Laser (COIL) has emerged as one of the most promising and powerful lasers in the recent past because of its immense potential in wide ranging applications in material processing and defense. The low temperature, low pressure COIL active medium provides low divergence and excellent beam quality. The wavelength of the laser (1.315 micron) is compatible for transmission through fiber and atmosphere. COIL system can be scaled up to several megawatts and can also be operated in a wide range of the pulse repetition regimes from Hz to GHz range. These features clearly indicate their potential in the industry for cutting and welding of materials, remote applications like cutting and dismantling of obsolete nuclear reactors or underwater cutting and in military applications as antimissile weapon.

Since its invention in 1978, the COIL research has passed through many stages and the system development has now reached to such a maturity level where advanced countries are planning deployable systems based on this laser. The COIL systems of hundreds of kilowatt power have been realized. Further efficiencies of the order of 35 % are becoming possible with the realization of supersonic flows in the cavity.

The article briefly presents the history of COIL along with the detailed review on design and development status of various critical subsystems like singlet oxygen generator, supersonic nozzle geometries, cavity configurations, and schemes for atmospheric discharge COIL. Various applications of COIL in civil and military including Airborne Laser, Tactical High Energy Laser, material processing, remote dismantling of nuclear reactors and rock crushing for oil exploration are also discussed.

Chapter 6 – 'Search for potential materials' for third-order nonlinear optics has been of continuing interest in recent years. In this context, organic molecules are increasingly being recognized as the materials of the future because their molecular nature combined with the versatility of synthetic chemistry can be used to alter and optimize molecular structure to maximize third order nonlinear response. Chalcones have received considerable interest as materials for second-order nonlinear optical applications due to their ability to crystallize in noncentrosymmetric structure and their blue light transmittance. Being charge transfer compounds, chalcones can also possess large third-order nonlinearities due to their π-conjugated structure. In this article, the authors discuss the structure-nonlinear response relationship among a few chalcones and the possibility of using them for third-order applications. A meager or no work, to their knowledge, has been done so far on these molecules in this regard. Z-scan and degenerate four-wave mixing techniques were employed to investigate third-order optical nonlinearities of chalcone derivatives. Some of these molecules possess large $\chi^{(3)}$ of magnitude as high as 10^{-12} esu and exhibit strong optical limiting properties. Possible mechanisms that are responsible for optical limiting property of these molecules have been discussed.

Chapter 7 – The authors report on optical analogues of well-known electronic phenomena such as Bloch oscillations and electrical Zener breakdown. The authors describe and detail the experimental observation of Bloch oscillations and resonant Zener tunneling of light waves in static and time-resolved transmission measurements performed on optical superlattices. Optical superlattices are formed by one-dimensional photonic structures (coupled microcavities) of high optical quality and are specifically designed to represent a tilted photonic crystal band. In the tilted bands condition, the miniband of degenerate cavity modes turns into an optical Wannier-Stark ladder (WSL). This allows an ultrashort light pulse to bounce between the tilted photonic band edges and hence to perform Bloch oscillations, the period of which is defined by the frequency separation of the WSL states. When the superlattice is designed such that two minibands are formed within the stop band, at a critical value of the tilt of photonic bands the two WSLs couple within the superlattice structure. This results in a formation of a resonant tunneling channel in the minigap region, where the light transmission boosts from 0.3% to over 43%. The latter case describes the resonant Zener tunneling of light waves.

Versions of chapters 1, 3, 4, 5 and 6 were also published in *Journal of Optics Research,* Volume 10, Numbers 1-4 published by Nova Science Publishers, Inc.; versions of chapters 2 and 7 were also published in *New Topics in Laser and Electro-Optics* edited by William T. Arkin published by Nova Science Publishers, Inc. They were submitted for appropriate modifications in an effort to encourage wider dissemination of research.

In: Perspectives in Optics Research
Editor: Jeffrey M. Ringer, pp. 1-17

ISBN: 978-1-61122-934-9
© 2011 Nova Science Publishers, Inc.

Chapter 1

RECENT PROGRESSES OF MULTI-WAVELENGTH SEMICONDUCTOR LASER DIODES FOR OPTICAL INFORMATION STORAGE SYSTEMS

H. Ko[1], M. W. Cho[2]

[1]Department of Electrical and Computer Engineering,
University of South Alabama, Mobile, AL, USA
[2]Institute for Materials Research, Tohoku University,
2-1-1 Katahira, Aoba-ku, Sendai, 980-8577, Japan

ABSTRACT

Due to the rapid technical advances of semiconductor laser diodes as a light source, the ultra-high-density optical information storage systems have been accelerated to commercialize in various formats, such as CD (Compact Disk), CD-ROM (Read-Only Memory), CD-R (-Recordable), CD-RW (-ReWriteable), DVD (Digital Versatile/Video Disc), DVD-R, and DVD-RW, for many applications, i.e., audio, music, image, movie, and computer data storage. In this complex situation, the compatibility of diverse storage media discs in a playback/writing system is one of the most important issues for system development. This problem has caused complications in the optical components as well as the non-compact system size. A semiconductor laser is an integral part among many optical components because the wavelength and output power of laser light source is a key factor of the media disc specifications.

In order to realize signal compact optical pick-up module that can respond to various media discs, a multiple wavelength laser light source is indispensable. There have been many attempts to develop multi-wavelength laser diodes. The typical approaches are to use two individually packaged laser diodes or hybrid-integrated two laser chips in the same package. However, those techniques require sophisticated assembly processing resulting in low reliability and high cost.

In this chapter, we reviewed the recent progress of multi-wavelength laser diodes and how it can effectively realize dual or multiple wavelength laser diodes in the future. We proposed, and successfully demonstrated, new concept device structures for monolithically-integrated two laser diodes with different wavelengths on a GaAs wafer in parallel and vertical direction.

1. INTRODUCTION

In the transitional period of an optical media disc, various kind of optical pick-up module has been developed for various formats, such as CD, CD-ROM, CD-R, CD-RW, DVD, DVD-R, and DVD-RW, for many applications, i.e., audio, music, image, movie, and computer data storage. Figure 1 shows the development trend for optical media disc since 1980's. During recent 5 years, the transition to short wavelength and high output power has drastically accelerated the progress of CD and DVD playback systems. The track pitch distance and minimum pit/land depth on a disc depend on the wavelength of LDs, and also are critical

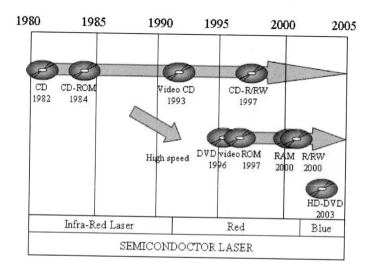

Figure 1. The development trend for optical media discs since 1980's.

	CD-R	DVD-R
Disc Diameter	80 mm, 120mm	80 mm, 120mm
Disc Thickness	1.2 mm	1.2 mm
Laser Wavelength	780 nm	635, 650 nm
Numerical Aperture	0.45	0.55, 0.60
Track Pitch	1.6 um	0.74 um
Shortest Pit /Land length	0.83	0.4 um
Data capacity	650 - 700 MB	1.50 GB and larger

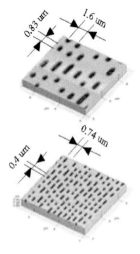

Figure 2. The comparison pf the disc specifications of CD-R and DVD-R.

factors to decide recording capacity. To increase of recording capacity, double sided structure disc was also developed. The recording capacity of optical storage media could be extended to ~ 8.5 Gbit in the double sided structure of one layer media disc by the excellent performance of the semiconductor laser diodes as a light source of a playback system.

Figure 2 shows the disc specifications of CD-R and DVD-R. A 650 or 635 nm laser diodes (LDs) and 780 nm LDs are used as light source for CD-R and DVD-R, respectively. In addition to wavelength, the access speed is mainly dependent on the output power of laser diode in the same wavelength. Recently, HD-DVD (High Density-Digital Versatile Disk) called "Blue-ray disc" as a new generation of storage media, which can record maximum up to 27 Gbit video data on a single sided phase change disc of 12 cm in diameter, has been realized by adopting a 405nm blue-violet GaN-based laser diode [1]. The performance improvement of LDs based on compound semiconductors has leaded a dramatic progress of optical media disc playback systems. All kinds of storage media discs, DVD-ROM, CD-R and CD-RW are immediately required to different optical pick-up heads. Unfortunately, one kind of light source cannot cover all optical media disc. For example, a 780 nm laser source must be added to a DVD pickup-head specially for reading CD-R's data because the reflectivity of CD-R media is lower than 10% for 650 nm light. In order to operate both DVD-R and CD-RW in one playback system, each optical pick-up head or multi-wavelength optical source corresponding to each disc has to exist within one playback system. Recently, a high-density optical disk system which has a capacity of more than 20 Gbit has been realized using a blue LD (450 nm of wavelength) as the light source of the optical pick up. In such system, a red LD (650 nm of wavelength) will still be need to playback the conventional DVD, DVD-R, DVD-RW, and DVD-RAM discs with the same system.

In this complex situation, the compatibility of diverse storage media discs in a playback/writing system is one of the most important issues for system development. This problem has caused complications in the optical components as well as the non-compact system size. A semiconductor laser is an integral part among many optical components because the wavelength and output power of laser light source is a key factor of the media disc specifications.

In order to realize single compact optical pick-up module that can respond to various media discs, a multiple wavelength laser light source is indispensable. There have been many attempts to develop multi-wavelength laser diodes. The typical approaches are to use two individually packaged laser diodes or hybrid-integrated two laser chips in the same package. However, those techniques require sophisticated assembly processing resulting in low reliability and high cost. The other approach is a monolithically-integrated two-wavelength laser diode on a GaAs wafer in parallel and vertical direction by multi-step epitaxial growth and sophisticated process. Therefore, a novel multiple-wavelength light emitter is strongly required to solve the fundamental problems and to improve the device performance.

In this chapter, we reviewed the recent progress of multi-wavelength laser diodes and how it can effectively realize dual or multiple wavelength laser diodes in the future. We proposed, and successfully demonstrated, new concept device structures for monolithically-integrated two laser diodes with different wavelengths on a GaAs wafer in parallel and vertical direction.

2. HYBRID-INTEGRATED TWO-WAVELENGTH LASER DIODES

2.1. Conventional Type Lasers

A conventional hybrid type two-wavelength laser diode (TWLD) where two distinct laser chips were mounted on a single stem were equipped with two optical paths to read out DVD and CD signals [2]. Two well-developed semiconductor laser diodes with different emitting wavelength are used in a conventional hybrid type. Although the hybrid type TWLD has the advantage of adopting optimized lasers and giving a wide selectivity of laser wavelength and output power, it has difficult in precise definition of the two laser emission spots that could increase the laser packaging cost. Consequently, the opto-mechanical pickup design becomes complicated and the volume of a head is enlarged.

There are two kinds of package type in the conventional hybrid TWLD as shown in fig. 3. Two laser chips are vertically or horizontally bonded on one stem. For the conventional hybrid TWLD, the most crucial issue is "How we can reduce the distance between two emitting spots". In the case of horizontal type, two chips were parallel arranged side by side on the same submount, however with a certain separation between the two emitting spots as shown in fig. 3 (a). In the case of vertical type, two chips were placed on the upper and lower step of the stem as shown in fig. 3 (b). The distance between the two emitting spots was primarily determined by the width of the chip and stem design. According to the stem design, the chip arrangement can be different from that mentioned above. Fig. 4 shows a scanning electron microscopy image of a horizontal type TWLD packaged with 650 and 780 nm LD. The emitting position of the two elements are 250 ~ 300 μm apart, which still requires separate optical alignment.

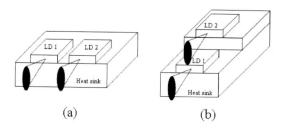

(a)   (b)

Figure 3. Two kinds of package type in the conventional hybrid TWLD. Two chips are (a) horizontally and (b) vertically bonded on the same stem.

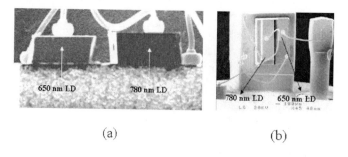

(a)   (b)

Figure 4. A SEM image of a horizontal type TWLD packaged with 650 and 7801 nm LD. (a) front side view and (b) top side view.

2.2. Holographic Pick-Up Module Type Lasers

In order to realize compact pick module, a holographic pick up-head module, which can package a light source, photo detectors, beam splitter, and servo generating optical elements, was suggested. Figure 5 shows typical two laser chips packaging for holographic pick-head module [3]. The laser chips are bonded on the upper and lower steps of the stem, respectively. The diachronic micro-mirror has two reflecting surfaces separated by 0.16 mm designed to compensate for the offset between the optical axes of the two beams. It is essentially a 45-degree tilted plate with thickness 0.16 mm bonded to a micro-mirror. The height of the submount is related to the separation of the two reflecting surfaces. The first surface has a wavelength select coating that reflects 650 nm laser light and transmits 780 nm light, while the second surface totally reflects 780 nm light. In the design of the stem and mirror, astigmatism that is introduced when the divergent laser beam passes through twice the 45 degree tilted plate should be considered to meet the strict optical requirement. It results in the advantages of cost reduction in packaging the component and the compactness of the unit. However, the construction of a module requires a mechanical alignment precision on the order of microns as well as allowance for wavelength shift of the laser diodes over several nanometers. To satisfy these strict alignment requirements, a robust focusing and tracking method have been being studied.

2.3. Chip-to-Chip Bonding Type Lasers

Though the conventional hybridized LD was attempted for compactness, it was needed to align separate optical paths corresponding to each LD because the distance between the two emitting spots is above a few hundred μm. In order to make the two emitting spots closer, we have vertically stacked two chips on the same stem. Fig. 6 shows the schematic diagram of a vertically hybridized chip-to-chip bonding package on a stem, and patterned solder to bond two chips. Three electrodes were formed, and p-electrode was common for two LDs.

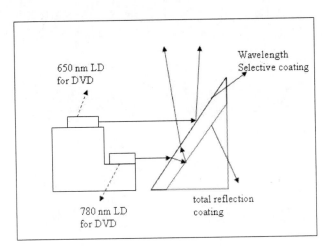

Figure 5. Holographic pick-head module packaging with two laser chips. The laser chips are bonded on the upper and lower steps of the stem, respectively.

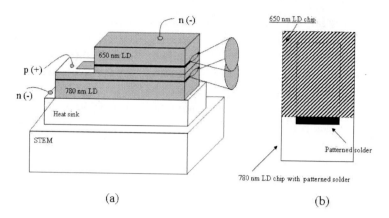

(a)   (b)

Figure 6. (a) The schematic drawing of a vertically hybridized chip-to-chip bonding on a stem. (b) Top side view of two LDs patterned with solder.

In this experimental, a high power 780 nm LD was used for high speed CD-RW, and a 650 nm LD was standard for DVD-R, respectively. It is the first attempt to make a CD-RW and DVD-R playback system with chip-to-chip bonding type. Fig. 7 shows the schematic diagram of the 650 and 780 nm LD structures, respectively. For 780 nm LD, the optimized graded index separate confinement heterostructure-multiple quantum well (GRINSCH-MQW) structure was used. The active layer was consisted of $Al_{0.3}GaAs/Al_{0.1}GaAs$ MQW, and the optical confinement layer changed the Al composition of AlGaAs layer from 32 % to 55 % within 516 Å. Epitaxial growth was carried out by low pressure MOCVD on an n-type (100)-5° off toward <011> n-GaAs substrate. A window mirror structure was adapted in the facet region so as to increase the catastrophic optical damage level. For 650 nm LD, AlGaInP-GRINSCH-MQW structure was used. The structure is a buried ridge stripe that is widely used in laser diodes applied to light sources in an optical disc system. An InGaP/AlGaInP MQWs and the AlGaInP layers were grown on n-type on a (100)-15° off toward <011> n-GaAs substrate as active and optical confinement layers, respectively.

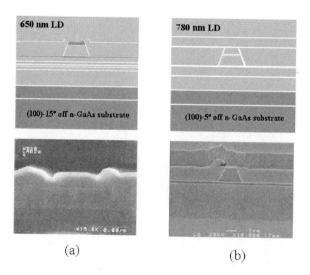

(a)   (b)

Figure 7. The schematic diagram and SEM images of (a) 650 and (b) 780 nm LD structures.

Table 1. Thermal conductivity of various materials.

	GaAs	GaN	Al$_2$O$_3$	AlN	Si
Thermal conductivity (W/cm K)	0.56	1.3	2.51	2.85	1.51

First, a high power 780 nm laser chip was junction-up bonded on a submount using a hard solder at high temperature. The chip size is 250 × 500 μm^2 and 250 × 900 μm^2 for 650 and 780 nm LDs, respectively. Then, a 650 nm LD is junction-down bonded on a 780 nm LD with the patterned soft solder at low temperature. Before packaging, hard solder metal was deposited on p-metal of 780 LD with 200 × 500 μm^2. Solder selection is the most important to bond two LDs without re-melted phenomenon of the hard solder. Three electrodes were formed with p-common. The distance between emitting spots can be minimized below a few μm apart because only solder and p-metal existed in interface between the two LDs. One optical path is distinguishable merit as compared with other TWLD LDs, and is an ultimate solution for TWLD.

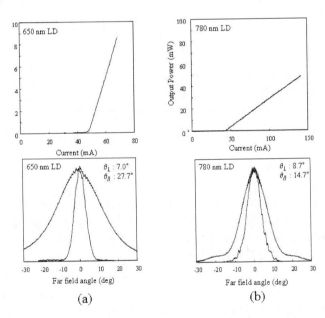

Figure 8. The output power versus current (*L-I*) characteristics and far field patterns for the (a) 650 nm LD and (b) 780 nm LD under continuous wave (cw) operation at room temperature after chip-to-chip bonding on a stem.

Figure 8 shows the output power versus current (*L-I*) characteristics and far field angles for the both LDs under continuous wave (cw) operation at room temperature after packaging. Two LDs were uncoated for both facets. The threshold current is 48 mA, maximum power (P$_0$) above 100 mW with kink-free operation for 780 LD. The threshold current is 43 mA, and maximum power (P$_0$) 10 mW for 650 LD. The beam divergence perpendicular to the junction at 5 mw and 100 mW was 27.7° and 14.7°, and that parallel to the junction was 7.0° and 8.7°, respectively. The cavity length is 250 × 500 μm^2 and 500 × 900 μm^2, respectively.

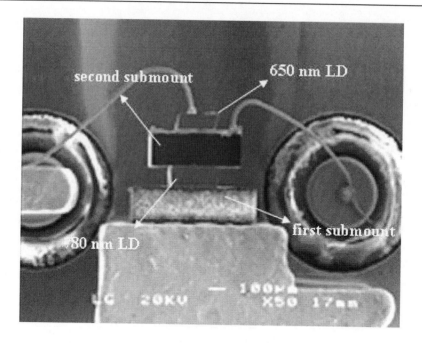

Figure 9. SEM image of chip-to-chip bonding packaged by the conductive type submount inserted between two LDs.

Basically, the 780 nm LD on the submount has not any problems for thermal dissipation due to junction-up bonding. But, in the case of the 650 nm LD, the thermal dissipation is a critical problem because it puts on the 780 LD of the thick ~ 80 μm. Generally, the thermal accumulated during operation of LD is one of the critical damages to device degradation. Although thermal conductivity of GaAs material is lower than that of the other heat-sink materials such as Si and AlN, GaAs LD can be expected as a role of heat sink as operating in the low power for DVD-R. The thermal of 650 LD during operation can flow through the 780 LD into heat sink because of metal solder with the high thermal conductivity in the interface between 780 LD and 650 LD. The heat sink material, the package type, the heat capacity of device, the metal system for p-electrode and the solder material much critically influence the thermal dissipation. Table 1 shows the thermal conductivity of various materials. In order to improve heat dissipation, we have attempted to insert the conductive type submount between two LDs as shown in fig. 9. Then, the generated heat can be effectively dissipated through the side edge of submount. Modifying the thickness of the submount can easily reduce the distance between the two emitting spots.

If the concept of chip-to-chip bonding technique applies to a blue GaN laser, the various combinations such as HD-DVD and DVD, HD-DVD and CD can be obtained. Generally, a GaN device was grown on sapphire substrate, and the thermal conductivity of GaN and Al_2O_3 substrate is higher than that of GaAs and the other submount materials. It is expected that GaN LD on Al_2O_3 totally acts on as a superior heat sink.

3. MONOLITHIC-INTEGRATED TWO WAVELENGTH LASER DIODES

The first monolithic type LD is originated from a laser diode array for high power operation above a few watts. However, it is not a light source for playback system of optical media disc but for applications like laser marker in the industrial or laser operation in the medical. Recently, a monolithic type two-wavelength laser diode on GaAs substrate is suggested for further reductions in cost and size of the optical disk playback system. TWLD combining two laser diodes is monolithically integrated on a single chip by photolithography processes and re-growth. The device structure and lasing characteristics of monolithic-type dual wavelength LD developed to date was summarized in the table 2.

3.1. Visible and Infrared Bands Integrated Type Lasers

Recently, Sony's research group reported a reliable monolithic-integrated two-wavelength visible and infrared laser diode structure (VILD) using two-step MOCVD growth [4]. The 650 nm element's epitaxial islands, and the 780 nm elements' epitaxial islands parallel to the <01-1> direction, are positioned side by side on a GaAs substrate as shown in

Table 2. Device structure and lasing characteristics of Monolithic TWLD.

Device structure of Monolithic TWLD	cavity length (μm²)	stripe width (μm)	facet coating	Emitting position (μm)	lasing characteristics			Ref.
					I_{th} (mA)	$\theta_\perp / \theta_\parallel$ (deg.)	Po (mW)	
(a) VILD	400 μm (650 LD) / 400 μm (780 LD)	6.5 (center) 3.5 (edge) (650 LD) / 8.0 (center) 5.0 (edge) (780 LD)	32 %/75 % (650 LD) / 28 %/75 % (780 LD)	120	~50 (650 LD) / ~55 (780 LD)	35.7/10.0 (650 LD) / 34.9/14.4 (780 LD)	low power (650 LD) / low power (780 LD)	[4]
(b) TWINLD	300 μm (650 LD) / 300 μm (780 LD)	3.5 (650 LD) / 2 (780 LD)	uncoated	150	~12 (650 LD) / ~14 (780 LD)	37/6 (650 LD) / 37/17 (780 LD)	low power (650 LD) / 35 (780 LD)	[5]
(c) RISA LD	800 μm (650 LD) / 800 μm (780 LD)		19 %/86 % (650 LD) / 10 %/96 % (780 LD)	120	~45 (650 LD) / ~25 (780 LD)	- (650 LD) / 17/8.2 (780 LD)	10 (650 LD) / 210 (780 LD)	[6,7]
ZnCdSe/ZnSe/ZnMgBeSe SCH-SQW / InGaP/InGaAlP SCH-MQW / GaAs 15°-off sub. (d) Green & red LD	300 μm (650 LD) / 600 μm (780 LD)		-	~3 μm	115 kW/cm² (650 LD) / 84 kW/cm² (780 LD)		-	[8]

table 2 (a). Both elements have a gain guiding structure fabricated by stripe pattern. A separated-confinement-heterostructure multi-quantum-well (SCH-MQW) structure is used for each active region to minimize the threshold current. The wave-guide stripe of the 650 nm element is a mesa-shaped tapered stripe, which forms a gain-guiding cavity. For the 780 nm element, a tapered-stripe gain-guiding structure is used, which is prepared using ion implantation. The length of the tapered stripe is 170 µm in both LDs. The stripe width is 6.5 µm at the center and 3.0 µm at the facet edge for the 650 nm LD, and 8.0 µm at the center and 5.0 µm at the facet edge for the 780 nm LD. The emitting positions of the two LDs are 120 µm apart. The laser chip is 120 mm thick and 300 mm wide. The cavity length is 400 µm. The reflectivity of the coated front and rear facets of for the 780 nm LD are 28 % and 75 % respectively, for the 650 nm LD 32 % and 75 % respectively. The waveguide stripes were formed using one-step photolithography, so the stripes for the 650 nm element and the 780 nm element were precisely positioned. Finally, after conventional processes, the laser chip is mounted on a Si-submount, which is mounted on a Cu heat sink in a conventional 5.6 mm φ package with four electrodes.

For the 650 and 780 nm LD, the operating currents at 5 mW were 57.0 and 61.5 mA, and kink-free operation was achieved up to 12 and 15 mW, respectively. The characteristics temperatures were estimated to be about 150 and 170 K, respectively. The beam divergence perpendicular to the junction at 5 mW was 35.7° and 34.9°, and parallel to the junction was 10.0° and 14.4°, respectively. Both elements showed stable fundamental transverse mode operation. The astigmatism was about 25 µm for both elements, which was a little smaller than that of a conventional gain-guiding laser with a straight stripe. The lasing spectrum of the 780 nm LD showed multimode emission with narrow linewidth longitudinal modes, which is typical for the tapered-stripe gain guiding laser diode that is widely used in the CD playback system. The lasing spectrum of the 650 nm LD was also multimode, and sometimes showed linewidth broadening of the longitudinal modes. The mean time to failure of the 650 and 780 nm elements was over 10,000 hr and over 100,000 hr from aging test under cw operation at 70 °C for the 650 and 780 nm elements at 6 and 7 mW, respectively. This performance allows VILD to support the 6X-speed DVD and 48X-speed CD playback system..

Although the reliable two-wavelength laser diode was realized using VILD structure, it has some restriction for various applications such as CD-RW and DVD-RW media that require high output power laser characteristics. In order to realize high power operation with stable fundamental transverse mode, it is critical to suppress spatial hole-burning in the mirror facet by reducing the operating current. Typically, the high power laser diode has adopted the index-guiding structure and facet coating process suitable to each wavelength for lowering threshold current and increasing the COD (catastrophic optical damage) level. The more complicated lithography and re-growth process were required for high power operation to achieve recording CD-R and DVD-RAM discs.

Recently, T. Lu et al. has also reported on TWINLD (Two-wavelength integrated laser diode), which has been realized by monolithically combining two different laser material structures in a single chip utilizing the MOCVD re-growth technique [5]. The TWINLD emits a red wavelength λ=650 nm and an infrared wavelength λ=780 nm for operation of DVD and CD, respectively. This is nearly the same structure as the VILD structure except for using aluminum-free DQW (double quantum well) in the active layer and a SiO_2 mask in the re-growth process as shown in table 2 (b). The TWINLD structure was remarkable for using an aluminum-free active area structure that could allow high-power operation for 780 nm LD. The 780 nm laser structure has an active region consisting of two –0.1% tensile strained $In_{0.08}Ga_{0.92}As_{0.83}P_{0.17}$ wells and one $Al_{0.3}Ga_{0.7}As$ barrier, which is adopted as p- and n-cladding layer with the same thickness as the cladding layers of the 650 nm laser. The first one is grown under the buffer layer in order to obtain a flat etched surface and to accurately control the depth of the re-growth area. And the second one is aimed for the p-type cladding layer to ensure the precision of etching depth when making a ridge wave-guide structure.

A 2 μm wide ridge for the 780 nm laser and a 3.5 μm wide ridge for the 650 nm laser were formed with a 150 μm separation between the two emission spots. From typical laser characteristics, the threshold current is 12 mA at 20 °C and the one end slope efficiency is 0.6 W/A for both facets uncoated for 650 nm laser. Similarly, for 780 nm lasers, the threshold current is 14 mA at 20 °C and the one end slope efficiency is 0.4 W/A for both facets uncoated for 780 nm lasers. The single ended output power under cw operation can reach up to 35 mW for both facets uncoated. Far field angles of 650 and 780 nm LDs are 6°/37° ($\theta_\square/\theta\perp$) and 17°/37° for the direction parallel and perpendicular to the junction plane, respectively. Although TWINLD structure shows much lower threshold current as compared with VILD structure, it is very closely related with ridge width and cavity length. Considering threshold current density (J_{th}) and aspect ratio, the device structure should be optimized. Finally, these factors give critical effect to device reliability.

Matsushita group reported a real refractive index guided self-aligned (RISA) structure to realize high power operation with low threshold current density [6, 7]. RISA structure is a promising optical waveguide structure for the reduction of the operating current because of negligible absorption loss in the current blocking and the built-in refractive index difference. Self-sustained pulsation (SSP) is advantageous for the reduction of additional components such as a quarter-wavelength phase plate or a high frequency superposition circuit in the optical head. In this means, SSP LDs are attractive devices as they can suppress optical feedback noise, due to the fact that mode competition can be suppressed by multi-longitudinal mode oscillation. Using RISA structure, they succeeded in developing SSP 650 nm band LD structure maintained at temperature up to 75°C at an average output power of 7 mW due to reduced current density. In order to obtain higher temperature SSP, a SA (saturable absorbing) layer, which was used for maintaining SSP operation under high power and high temperature, is incorporated in the 650 nm band LD.

The 780 nm AlGaAs and 650 nm AlGaInP lasers are monolithically integrated on a GaAs substrate as shown in table 2 (c). The RISA waveguide structure is adopted for both LDs. The 780 LD has an AlGaAs bulk active layer and an AlGaAs current blocking layer transparent in the lasing wavelength. The 650 nm LD has GaInP/AlGaInP compressively strained multiple quantum wells as the active region, and an AlInP current blocking layer, which are also transparent in the lasing wavelength. The LDs are fabricated by four-step MOCVD growth. The distance between the emitting points of 780- and 650-nm band LDs is set to be 120 μm, and cavity length is 800 μm. The front and rear facets are coated with reflectivity of 10 and 96% for the wavelength of 780 nm, respectively. For the wavelength of 650 nm, the reflectivity of each facet is 19 and 86 %, respectively. As the material with high refractive index for the high reflectivity coating on the rear facet, TiO_2 films are selected in order to avoid optical loss and subsequent heat generation in the coating films.

For 780 nm LD, the maximum output power over 200 mW is attained and the operating current is as low as 240 mA at 200 mW under cw operation. Such a high output power and low operating current at a high power level are realized due to reduced internal loss in RISA structure. No kinks are observed up to the COD level, which indicates that spatial hole-burning is effectively suppressed. For 780 nm band LD, beam divergence angles perpendicular and parallel to the junction plane are 17° and 8.2°, respectively, resulting in a smallest aspect ratio of 2.1 as compared with the other TWLDs. It is shown that the RISA structure is highly effective for improving the SSP characteristics in monolithically integrated dual wavelength LDs. Moreover, it means that high power operation based on index-guiding type on monolithic laser chip is applicable to CD-RW media.

3.2. Green and Red Bands Integrated Type Lasers

A monolithic type dual wavelength developed up to date has inherent problems such as wavelength selectivity limited to GaAs-based material and the distance between the emitting points between two LDs due to horizontally integrated arrangement. In order to overcome the fundamental problems, we have attempted for realizing vertically integrated ZnSe-based II-VI material grown on InGaP-based III-V structure as shown in table 2 [d] [8].

The merits of this structure are as follows: 1) lattice matching epitaxial growth could be achieved both III-V and II-VI materials on a GaAs substrate, 2) the band gap of these material systems covers wide range from blue to infrared, therefore various combination is possible, 3) simple packaging will be possible as a result of monolithic integration because emitting light will be placed in a few μm distances, which will greatly simplify the optics design, 4) in the point view of epitaxial growth, thermal damage of III-V laser structure due to second MBE growth can be ignored, because II-VI growth temperature of MBE is considerably lower than III-V of MOCVD.

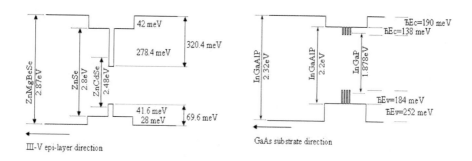

Figure 10. (a) band alignment for II-VI blue-green section (b) band alignment for III-V red section.

However, InGaP epitaxial layers grown on (001) GaAs substrate by MOCVD has an ordering problem. It is well known that (001) GaAs substrate misoriented toward [110] prevent ordering in the InGaP/GaAs system [9, 10]. On the other hand, there are some disadvantages for MBE growth of ZnSe-base II-VI on misoriented GaAs substrate, since the surface of the misoriented toward [110] GaAs substrate is covered with Ga-step. It is hard to grow ZnSe epitaxial layer on this substrate by MBE due to increased possibility of reaction with Ga and Se.

Since the ZnSe-based light emitting devices have been fabricated by MBE on GaAs (001) substrates, one of the important growth issues would be the growth of high-quality ZnSe-related materials on tilted GaAs (001) substrates. GaAs buffer layers, the pretreatment of Zn exposure on the GaAs in the initial growth stage, and low-temperature-grown ZnSe (LT-ZnSe) buffer have been adopted. In particular, LT-ZnSe buffer layer and the pretreatment of Zn exposure are introduced to obtain a smooth surface and to facilitate high temperature growth of ZnSe layers.[13] We succeeded in improving ZnSe crystallinity by inserting low temperature (LT)-ZnSe buffer between the high temperature (HT)-ZnSe epitaxial layer and the GaAs buffer on GaAs substrate. LT-ZnSe buffer was grown to reduce the formation of Ga-Se bonding, a well-known source of defect generation [11], due to inter-diffusion through the hetero-interface in the initial stage of growth. The stacking fault density of HT-ZnSe layer is considerably reduced to $2 \times 10^6/cm^2$ by inserting a thin LT-ZnSe buffer layer in the ZnSe/GaAs interface.

In order to show the feasibility of one-chip-two-wavelength light emitting device structure with vertical line-up of MBE grown ZnSe-based II-VI structures on MOCVD grown InGaP-based III-V structures, the SCH-SQW ZnCdSe/ZnSe/ZnMgBeSe structure is grown by MBE on the SCH-MQW InGaP/InGaAlP device structure which is grown on a GaAs (001) substrate tilted by 15° toward [110]. Two-step epitaxy growth is adopted for the one-chip-multiple-wavelength LDs. The MOCVD growth condition of the InGaP/InGaAlP structure was reported previously in detail [12, 13]. Based on such performances in ZnSe layers grown on (001) GaAs substrate tilted by 15° toward [110] direction, high-quality ZnCdSe/ZnSe/ZnMgBeSe structures were fabricated. Based on the previous reports of the band offsets in a ZnCdSe/ZnSe [14] and ZnSe/ZnMgBeSe QW structure [15], the band line up for the ZnCdSe/ZnSe/ZnMgBeSe QW is derived as shown in fig. 10 (a). The band offsets for the conduction and valence bands at the ZnCdSe/ZnSe interface are roughly estimated to be 278.4 meV and 41.6 meV, while those at the ZnSe/ZnMgBeSe are 42 meV and 28 meV. Although the valence band offset of 41.6 meV at the ZnCdSe/ZnSe interface is small to achieve enough carrier confinement at room temperature, the valence band offset of 28 meV at the ZnSe/ZnMgBeSe combined with the ZnCdSe/ZnSe band offset can help carrier confinement. The conduction band offset at the ZnCdSe/ZnSe interface is 278.4 meV, while 42 meV at the ZnSe/ZnMgBeSe. On the other hand, the band line up for the InGaP/InGaAlP QW structure is shown in fig. 10 (b) [16]. The band offsets for the conduction and valence bands of the InGaP/InGaAlP are 190 meV and 252 meV.

The degradation mechanisms of II-VI laser diodes based on sulfur containing II-VI materials have been studied in detail over the past years [17]. The relative abundance of lattice defects like dislocations, point defects, defect complexes in II-VI compounds, which lead to device degradation and limited lifetime, has been attributed to the relatively large ionicity of these materials. Beryllium chalcogenides are somewhat special II-VI materials with respect to lattice rigidity, which is due to their pronounced covalent bonding, in contrast to the much more ionic bonding counterparts like, e.g., ZnSe, and CdTe. Beyond the fundamental interest in these new materials, the increased bond strength can be expected to have an impact on the defect generation and propagation, and therefore the lifetime, of light emitting devices based on II-VI compounds.

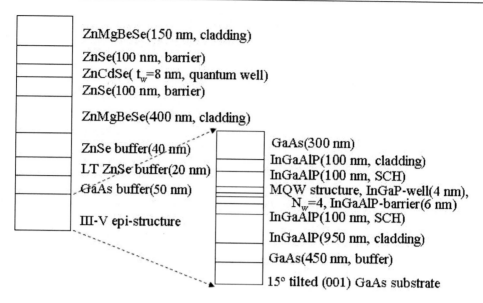

Figure 11. The schematic drawing of the QW structure of the multiple wavelengths light emitter for optical pumping experiments.

We performed optical-pumping experiments to confirm the feasibility of our proposed structure as a new candidate for the multiple-wavelength LDs. Specimens were mechanically polished to a thickness of about 80 μm and cleaved for optical pumping experiments with various cavity lengths. A frequency-doubled, tripled Nd:YAG laser (532 nm, 355 nm) was used as an excitation source for optical pumping experiments, III-V structure is optically pumped with 532 nm and II-VI structure is pumped with 355 nm excitation light source, respectively. The excitation light was irradiated normal to the sample surface and the emission was detected from one of the cleaved facets. Figure 11 shows the schematic diagrams of the QW structure of the multiple wavelengths light emitter for optical pumping experiments. ZnSe layers are inserted in between the ZnCdSe QW layer and ZnMgBeSe cladding layers to minimize interface fluctuations. The QW structures of the multiple wavelength light emitter consist of 8 nm thick ZnCdSe QW layer, 100 nm thick ZnSe act as barrier and SCH layer, 150 nm thick upper ZnMgBeSe and 400 nm thick lower ZnMgBeSe cladding layer in the blue-green region, while 4 nm thick InGaP QWs, 6 nm thick InGaAlP barriers, 100 nm thick InGaAlP SCH layers and 100 nm thick upper InGaAlP and 950 nm thick lower InGaAlP cladding layers in the red emitting III-V region.

Figure 12 (a) shows the evolution of edge emission spectra measured from the cleaved facet of the II-VI region with a cavity length of 300μm. Lasing occurred at the wavelength of 504 nm above threshold at the room temperature, as can be seen from the clear threshold behavior. The threshold power for lasing is estimated at 115 KW/cm^2. The emission spectra from the III-V region with a cavity length of 600 μm are shown in fig. 12 (b). The lasing wavelengths and threshold power are 664 nm, and 84 KW/cm^2, respectively. The combination of 504 nm and 664 nm wavelength could emit white color light, color temperature in the International Commission on Illumination (CIE) chromaticity coordinates is 3000 K. Longitudinal modes and polarization dependence for the II-VI section are observed as shown in fig. 13 (a). The mode spacing is estimated to be 1.6 Å. The cavity length is calculated as 300 μm using the relation $\Delta\lambda = \lambda^2/2nL$, where $\Delta\lambda$, L, n and λ are the mode spacing, cavity length, refractive index, and wavelength, respectively. We used the refractive index of 2.69 for ZnCdSe, which is approximated by the value of ZnSe. It does not make any difference due to small Cd composition of about 23%. This difference is within the resolution of our high excitation PL system. This value agrees with the measured cavity length of 300 μm by optical microscopy. As an additional evidence for lasing, we have measured the polarization properties of the emission above threshold. The emitted light shows strong transverse-electric (TE) polarization mode against transverse-magnetic (TM) mode with a polarization ratio of 29:1 (TE:TM). The angle of 90° means the emitted TE mode light and analyzer has the same direction. And 180° represent the emitted TM mode light and analyzer has the same direction. It strongly indicates that the electron to heavy hole transition is dominant in the quantum well, which corresponds to the compressive strain regime in the ZnCdSe quantum well. In fig. 13 (b), mode spacing for the III-V section are shown, the mode spacing is estimated to be 1.0 Å at the cavity length of 600 μm. We used the refractive index of 3.625 for InGaP.

This new approach has clearly shown the feasibility of one-chip integration of multiple-wavelength LDs with vertically integrated structure of II-VI and III-V compounds. The new integration technology of ZnSe-based blue-green light emitter grown by MBE on InGaP/InGaAlP red light emitter grown by MOCVD will be useful for a variety of application fields.

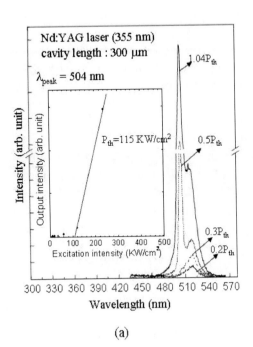

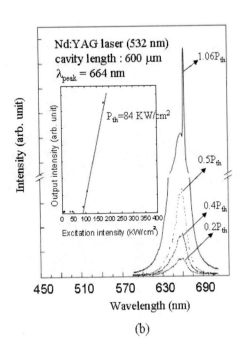

(a)

(b)

Figure 12. (a) The edge emission spectra measured from the cleaved facet of the II-VI region with a cavity length of 300 μm. (b) The emission spectra from the III-V region with a cavity length of 600 μm.

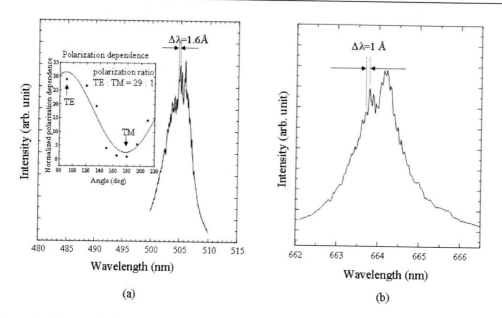

Figure 13. (a) Longitudinal maodes and polarization dependence for the II-VI section. (b) Mode spacing for the III-V section.

4. CONCLUSION

The Hybrid integrated two-wavelength laser diode is a desirable approach for compact optical pick-ups, since the simple package process and wide selectivity of wavelength and output power is to be possible. But, the distance between two emitting spots is still problem. To reduce the spot distance below a few μm, we have suggested vertically stacking chip-to-chip bonding technique, and it could make the spot distance very close below a few μm. But, the thermal dissipation of the upper laser has emerged as a new problem.

A monolithically integrated type two-wavelength laser diode has required high epitaxial growth technique and complicated process. Also, it was limited to the wavelength selectivity based on GaAs material, and very difficult to realize the high power operation of red laser diode on monolithic TWLD. Recent commercialized high-speed RW 52× CD and 16× DVD system require high power operation above ~240 mW. New index-guide laser structure with device performance equivalent to single high power LD chip have been required for stable high power operation of monolithic TWLD.

With appearance of "Blue ray disc" based on GaN material, it is expected that the whole market of the optical playback system may be moved to DVD or HD-DVD in the near future. In the present epitaxial growth technique, it is very difficult to monolithically combine two or three LDs based on GaAs and GaN materials. To hybridize blue and red or infrared lasers, new concepts like new package technique or wafer fusion will be needed. In this means, the chip-to-chip bonding technique is the most realizable technique in combining GaN-based blue laser and GaAs-based red or infrared laser, and then both the distance between two emitting spots and thermal dissipation can be solved due to GaN on Al_2O_3 with high conductivity and the small gap between two laser chips.

REFERENCES

[1] Ichimura, I.; Maeda, F.; Osato, K.; Yamamoto, K.; Kasami Y. *Jpn. J. Appl. Phys.* P1 2000, 39, 937-940.

[2] Lee, C.W.; Chung, C.S.; Yoo, J.H.; Lee, Y.H., Kim, T.K., Son, Y.K.; Kim; S.J.; Seong, P.Y.; Kim, K.S. *Jpn. J. Appl. Phys.* P1 1998, 37, 2197-2200.

[3] Shih, H.; Yang, T.; Freeman, M.; Wang, J.; Yau, H.; Huang, D. *Jpn. J. Appl. Phys.* P1 1999, 38, 1750-1754.

[4] Nemoto, K.; Kamei, T.; Abe, H.; Imanishi, D.; Narui, H.; Hirata, S. *Appl. Phys. Lett.* 2001, 78, 2270-2272.

[5] Lu, T.; Fu, R.; Shieh, H.M.; Huang, K.J.; Wang, S.C. *Appl. Phys. Lett.* 2001, 78, 853-856.

[6] Onishi, T.; Imafuji, O.; Fukuhisa, T.; Mochida, A.; Kobayashi, Y.; Yuri, M.; Itoh, K.; Shimizu, H. *IEEE Phot. Tech. Lett.* 2001, 13, 550-552.

[7] Onishi, T.; Imafuji, O.; Fukuhisa, T.; Mochida, A.; Kobayashi, Y.; Yuri, M.; Itoh, K.; Shimizu, H. *Jpn. J. Appl. Phys.* P1 2001, 40, 6401-6405.

[8] Song, J.S.; Cho, M.W.; Oh, D.C.; Makino, H.; Hanada, T.; Yao, T.; Zhang, B. P.; Segawa, Y.; Chang, J.H.; Song, H.S.; Cho, I.S.; Kim, H.W.; Jung, J.J. *Appl. Phys. Lett.* 2003, 82, 4095-4097.

[9] Gomyo, A.; Kawata, S.; Suzuki, T.; Iijima, S.; Hino, I. *Jpn. J. Appl. Phys.* 1989, 28, L1728-L1730.

[10] Gomyo, A.; Suzuki, T.; Kobayashi, K.; Kawata, S.; Hino, I. *Appl. Phys. Lett.* 1987, 50, 673-675.

[11] Kuo, L. H., Kimura, K., Ohtake, A., Miwa, S., Yasuda, T., Yao, T. 1997, *J. Vac. Sci. Tech.* B, 15, 1241-1253.

[12] Choi, W.J.; Kim, J.S.; Ko, H.C.; Chung, K.W.; Yoo, T.K. *J. Appl. Phys.* 1995, 77, 3111-3114.

[13] Choi, W.J.; Chang, J.H.; Choi, W.T.; Kim, S.H.; Leem, S.J.; Yoo, T.K. *IEEE J. Selected Topics in Quantum Electronics.* 1995, 1, 717-720.

[14] Guenaud, C.; Deleporte, E.; Filoramo, A.; Lelong, Ph.; Delalande, C.; Morhain, C.; Tournie, E.; Faurie, J. P. *J. Cryst. Growth* 1998, 184/185, 839-843.

[15] Chang, J.H.; Cho, M.W.; Godo, K.; Makino, H.; Yao, T. *Appl. Phys. Lett.* 1999, 75, 894-896.

[16] Tanaka, H., Kawamura, Y.,; Nojima, S.; Wakita, K.; Asahi, H. J. *Appl. Phys.* 1987,61, 1713-1719.

[17] Gua, S.; de Puydt, J.M.; Haase, M.A.; Qui, J.; Cheng, H *Appl. Phys. Lett.* 1993, 63, 3107-3109.

In: Perspectives in Optics Research
Editor: Jeffrey M. Ringer, pp. 19–30

ISBN: 978-1-61122-934-9
© 2011 Nova Science Publishers, Inc.

Chapter 2

SOLITON PROPAGATION IN THREE-LEVEL ATOMIC SYSTEM UNDER DETUNED EXCITATION

N. Boutabba[1][*] and H. Eleuch[1,2][†]
[1] Institut National des Sciences Appliquées
et de Technologie, BP : 676,
Zone Urbaine Nord 1080, Tunis, Tunisia
[2] International Center of Theoretical Physics,
Trieste, Italy

Abstract

We analyze the propagation of soliton pulses in an absorbing non-linear three-level medium in lambda configuration. One of the two atomic transitions is excited by a strong CW control laser light, whereas the other atomic transition is pumped by a weak variable light field. We take into consideration the effects of detunings between the laser sources and the two atomic transitions. Assuming two-photon resonance, we present an analytical expression of soliton shape. The soliton propagation velocity is influenced by three parameters: the amplitude of the stationary control field, the maximum amplitude of the soliton and by the detuning. Especially, we show that for a given value of the detuning, the soliton can be stopped.

1. Introduction

Solitons represent a fascinating aspect of the nonlinear phenomena [1] [2]. They are particular waves propagating through media without spreading and have particle-like properties. Recently, solitons have found applications in various fields such as plasmas physics [3], solid state physics [4], biological systems [5], shock waves , acoustic waves [6], electrical transmission lines [7], non-linear optics [8], chemical and geological systems [1].

The discovery of Electromagnetically Induced Transparency (EIT) led to the occurrence of new effects and new techniques including ultraslow light pulse propagation [9], the light signal storage [10] in atomic vapors or in an " atomic crystal" [11], the cooling of ground

---

[*]E-mail address: n_boutabba@yahoo.fr
[†]E-mail address: hichemeleuch@yahoo.fr

state atoms and the giant cross-Kerr non- linearity [12]. The phenomenon of electromagnetically induced transparency (EIT) as the propagation of laser radiation through a medium
is investigated and explained in terms of quantum coherence and interference for three-
level atomic systems . In the Λ configuration , a pair of optical pulses propagate without
absorption .This medium can be made experimentally [14] if two lasers are applied to a
resonant three-level system, the atoms will be driven to a population trapped state, and a
medium that is opaque to a probe laser can, by applying both lasers simultaneously, be made
transparent [15], [16]. This is especially of importance in spectroscopic applications, since
under these conditions spontaneous emission from the upper level is completely absent.

In our work we present a theoretical investigation of light pulse propagation in an atomic
ensemble consisted of 3-level system in Λ configuration. One of the two atomic transitions
is excited by a strong CW control laser light, whereas the other atomic transition is pumped
by a weak variable light field. We neglect the boundary condition problems and we use a
semi-classical approach with two classical monomode electromagnetic fields. Mainly, our
model take into account the dissipations via radiative emissions and the detuning effects.

The analytical expression of soliton's shape shows that the pulse propagation velocity
depends on the detuning between the two atomic transitions and the light sources, on the
control field amplitude and on the amplitude of the soliton.

In the next section of this paper, we outline the essential features of our theoretical
model, we describe the Λ-medium and the basic equations for its dynamical behavior. Section 3 is devoted to the study of the analytical soliton shapes in this medium. In Section 4,
we deduce the expression of the soliton velocity and we study its proprieties.

2.   Model

Let us consider an absorbing three level system in lambda configuration (three-level atom)
interacting with two resonant electromagnetic field. The medium is excited by a stationary
strong laser field (cw-field) applied on the stokes transition and the second laser field with
dynamical amplitude excites the pump transition (figure 1).

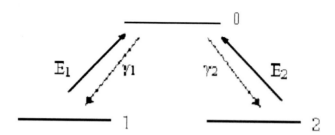

Figure 1. The three-level system in the lambda configuration

The three-level atom is described by quantum system with three energy levels $|0\rangle$, $|1\rangle$
and $|2\rangle$. The transitions $|0\rangle \leftrightarrow |1\rangle$ and $|0\rangle \leftrightarrow |2\rangle$ are possible whereas the levels $|1\rangle$ and $|2\rangle$
are supposed to be decoupled (transition $|1\rangle \leftrightarrow |2\rangle$ negligible). The reason of the choice of
this model is the fact that a free atom has at least two states at same parity between which
an electrical dipole transition is not allowed. The restriction to two lower energy level is

valid if the frequency of the interacting waves are distant enough to all other frequencies. In this model we take into account the rates $\gamma_{1,2}$ of radiative decay from the higher level $|0\rangle$ to the levels $|1\rangle$ and $|2\rangle$ and neglecting the other dissipation effects.

This three-level system is irradiated by a light beam propagating along an arbitrary direction x, with polarization adequate to couple the two optical transitions, and containing two monochromatic fields. This light beam is classically described as follows:

$$E'(x,t) = E_1(x,t) + E_2(x,t) = \bar{E}_1(x,t)\exp(-i\omega_1 t) + \bar{E}_2(x,t)\exp(-i\omega_2 t) \quad (1)$$

$\bar{E}_1$ and $\bar{E}_2$ are the amplitudes of the two waves. $\bar{E}_1$ is the amplitude of the cw-field and $\bar{E}_2$ is assumed to be slowly varying function in the following sense [17], [18], [19]:

$$\frac{1}{\omega_2}\left|\frac{\partial \bar{E}_2}{\partial t}\right| << \left|\bar{E}_2\right| \quad (2)$$

The Hamiltonian describing the interaction of the three level atom with the field has the expression :

$$H = H_0 + H_1 + H_2 \quad (3)$$

The first term of the Hamiltonian H_0 corresponds to the proper energies of the atom :

$$H_0 = \sum_{i=0}^{2} \varepsilon_i a_i^+ a_i \quad (4)$$

Where a_i, a_i^+ are respectively the annihilation and creation fermions operators of the atomic level i and ε_i the energy of the levels i. a_i, a_i^+ verifying the anticommutation relation:

$$\left[a_i, a_j^+\right]_+ = \delta_{ij} \quad (5)$$

The second (H_1) and third (H_2) terms of the Hamiltonian describe the interaction between the two fields and the atom[2]:

$$H_1 = g_1\left(a_0^+ a_1 E_1 + a_1^+ a_0 E_1^*\right)$$

$$H_2 = g_2\left(a_0^+ a_2 E_2 + a_2^+ a_0 E_2^*\right) \quad (6)$$

The two dipole transition matrix elements which are assumed to be real are denoted by g_1 and g_2.

To study the evolution of interaction between the atom and fields we use the density matrix formalism. The three level density matrix equation of motion is

$$\frac{d}{dt}\rho = \frac{1}{i\hbar}[H, \rho] + \frac{d}{dt}\rho_{irr} \quad (7)$$

where $\frac{d}{dt}\rho_{irr}$ describes the dissipation in the total system, and can be modelled by inter-action of the total system with thermal reservoir at the temperature 0 K of the thermal bath. $\frac{d}{dt}\rho_{irr}$ has the following expression :

$$\frac{d}{dt}\rho_{irr} = \frac{\gamma_1}{2}\left[a_0^+ a_1 \rho, a_1^+ a_0\right] + \frac{\gamma_1}{2}\left[a_0^+ a_1, \rho, a_1^+ a_0\right] + \frac{\gamma_2}{2}\left[a_0^+ a_2 \rho, a_2^+ a_0\right] + \frac{\gamma_2}{2}\left[a_0^+ a_2, \rho, a_2^+ a_0\right] \tag{8}$$

In term of the basis set of the bare atom $\{|0\rangle, |1\rangle, |2\rangle\}$ we have :

$$a_i^+ a_j = |i\rangle \langle j|$$

$$\rho = \sum_{i,j=0}^{2} |i\rangle \rho_{ij} \langle j| \tag{9}$$

The expression of the total Hamiltonian of the system becomes, in this basis:

$$H = \sum_{i=0}^{2} \varepsilon_i |i\rangle \langle i| + g_1\left(E_1 |0\rangle \langle 1| + |1\rangle \langle 0| E_1^*\right) + g_2\left(E_2 |0\rangle \langle 2| + |2\rangle \langle 0| E_2^*\right) \tag{10}$$

In order to derive the master equation (8), we are held to use the following intermediate calculations of the commutators:

$$[H_0, \rho] = \sum_{i'=0}^{2}\sum_{i,j=0}^{2} \varepsilon_{i'}\rho_{ij}\left(\delta_{ii'}|i'\rangle \langle j| - \delta_{ji'}|i\rangle \langle i'|\right) \tag{11}$$

$$[H_1, \rho] = \sum_{i,j=0}^{2} \rho_{ij}g_1 E_1\left(\delta_{1i}|0\rangle \langle j| - \delta_{0j}|i\rangle \langle 1|\right) + \sum_{i,j=0}^{2} \rho_{ij}g_1 E_1^*\left(\delta_{0i}|1\rangle \langle j| - \delta_{1j}|i\rangle \langle 0|\right) \tag{12}$$

$$[H_1, \rho] = \sum_{i,j=0}^{2} \rho_{ij}g_2 E_2\left(\delta_{2i}|0\rangle \langle j| - \delta_{0j}|i\rangle \langle 2|\right) + \sum_{i,j=0}^{2} \rho_{ij}g_2 E_2^*\left(\delta_{0i}|2\rangle \langle j| - \delta_{2j}|i\rangle \langle 0|\right) \tag{13}$$

$$[a_0^+ a_{i'}\rho, a_{i'}^+ a_0] = \sum_{i,j=0}^{2} \delta_{ii'}\rho_{ij}\left(\delta_{ji'}|0\rangle \langle 0| - |i'\rangle \langle j|\right) \tag{14}$$

$$[a_0^+ a_{i'}, \rho a_{i'}^+ a_0] = \sum_{i,j=0}^{2} \delta_{ji'} \rho_{ij} \left(\delta_{ii'} |0\rangle \langle 0| - |i\rangle \langle i'| \right) \tag{15}$$

So, we obtain the motion equation of the density matrix:

$$\frac{d}{dt}\varrho_{10} = i\,\omega_{10}\varrho_{10} - i\,d_1 E_1^* (\rho_{00} - \rho_{11}) + id_2\varrho_{21}^* E_2^* - \left(\frac{\gamma_1 + \gamma_2}{2} \right) \varrho_{10}$$

$$\frac{d}{dt}\varrho_{20} = i\,\omega_{20}\varrho_{20} - i\,d_2 E_2^* (\rho_{00} - \rho_{22}) + id_1\varrho_{21} E_1^* - \left(\frac{\gamma_1 + \gamma_2}{2} \right) \varrho_{20}$$

$$\frac{d}{dt}\varrho_{21} = i\,d_1 E_1 \rho_{20} + id_2\varrho_{10}^* E_2^* + i\,(\omega_{20} - \omega_{10})\,\varrho_{00}$$

$$\frac{d}{dt}\varrho_{jj} = i\,d_j \left(E_j \rho_{j0} - E_j^* \rho_{j0}^* \right) + \gamma_j\varrho_{00} \quad for\ j = 1, 2 \tag{16}$$

$$\frac{d}{dt}\varrho_{22} = i\,d_2 \left(E_2\rho_{20} - E_2^* \rho_{20}^* \right) + \gamma_2\varrho_{00}$$

$$\frac{d}{dt}\varrho_{ij} = \frac{d}{dt} \left(\varrho_{ji} \right)^* \quad for\ i, j = 0, 1, 2 \tag{17}$$

where ω_{10} and ω_{20} are the two atomic transition frequencies :

$$\omega_{10} = \frac{\varepsilon_0 - \varepsilon_1}{\hbar}$$

$$\omega_{20} = \frac{\varepsilon_0 - \varepsilon_2}{\hbar} \tag{18}$$

and $d_i = \frac{g_i}{\hbar}$ are the coupling constants.

From the property of the density matrix $tr\,(\rho) = 1$ we obtain:

$$\frac{d}{dt}\varrho_{00} = -\frac{d}{dt} \left(\varrho_{11} + \rho_{22} \right) \tag{19}$$

The diagonal elements of the density matrix ϱ describe the level populations and determine the internal energy of the atom. The off-diagonal elements describe the atomic coherences. The ϱ_{10} and ρ_{20} terms oscillate at the respective driving field frequency and the ϱ_{21} oscillate with frequency differences of the two light fields. So, we can define the slowly varying amplitudes of the off-diagonal density matrix elements ρ_{10}^-, ρ_{20}^- and ρ_{21}^- through the relations:

$$\rho_{j0} = \rho_{j0}^- \exp(i\omega_{j0}t) \quad for\ j = 1, 2$$

$$\rho_{21} = \rho_{21}^- \exp(i\,(\omega_{20} - \omega_{10})\,t) \tag{20}$$

We decompose the off-diagonal elements into an imaginary part and a real part :

$$\rho_{j0}^- = \chi_{j0} + i\psi_{j0}$$

$$\rho_{21}^- = \chi_{21} + i\psi_{21} \qquad (21)$$

The Hermitian propriety of the density matrix ensures that the diagonal elements ρ_{11}, ρ_{22} and ρ_{00} must be real. The evolution equations (8) become:

$$\frac{d}{dt}\chi_{10} = -\delta_1\psi_{10} + d_2\psi_{21}\bar{E}_2 - \left(\frac{\gamma_1 + \gamma_2}{2}\right)\chi_{10}$$

$$\frac{d}{dt}\psi_{10} = -\delta_1\chi_{10} - d_1\bar{E}_1\left(\rho_{00} - \rho_{11}\right) + d_2\chi_{21}\bar{E}_2 - \left(\frac{\gamma_1 + \gamma_2}{2}\right)\psi_{10}$$

$$\frac{d}{dt}\chi_{20} = -\delta_2\psi_{20} - d_1\psi_{21}\bar{E}_1 - \left(\frac{\gamma_1 + \gamma_2}{2}\right)\chi_{20}$$

$$\frac{d}{dt}\psi_{20} = \delta_2\chi_{20} - d_2\bar{E}_2\left(\rho_{00} - \rho_{22}\right) + d_1\chi_{21}\bar{E}_1 - \left(\frac{\gamma_1 + \gamma_2}{2}\right)\psi_{20} \qquad (22)$$

$$\frac{d}{dt}\chi_{21} = -d_1\bar{E}_1\psi_{20} - d_2\psi_{10}\bar{E}_2 - (\delta_2 - \delta_1)\psi_{21}$$

$$\frac{d}{dt}\psi_{21} = d_1\bar{E}_1\chi_{20} - d_2\chi_{10}\bar{E}_2 + (\delta_2 - \delta_1)\chi_{21}$$

$$\frac{d}{dt}\rho_{11} = -2\,d_1\bar{E}_1\psi_{10} + \gamma_1\rho_{00}$$

$$\frac{d}{dt}\rho_{22} = -2\,d_2\bar{E}_2\psi_{20} + \gamma_2\rho_{00}$$

$$\frac{d}{dt}\varrho_{00} = -\frac{d}{dt}\left(\varrho_{11} + \rho_{22}\right)$$

where δ_1 and δ_1 are the detunings between the laser frequencies and the atomic transitions frequencies:

$$\begin{aligned}\delta_1 &= \omega_{10} - \omega_1 \\ \delta_2 &= \omega_{20} - \omega_2\end{aligned} \qquad (23)$$

In this section, we have developed the evolution of the atomic parameters. In the next section, we study the propagation of the field E_2 through the medium: more precisely, we explore its spatial and temporal dynamical behavior.

3.   Soliton in the Three-Level Medium Pumped by CW-field

This section deals with the analysis of soliton propagation in the medium described above. We focus on the case of a strong stationary control field applied to the transition $|0\rangle \leftrightarrow |1\rangle$ and a much weaker signal field at the transition $|0\rangle \leftrightarrow |2\rangle$. The signal field E_2 is described by the Maxwell equation for a slowly varying approximation (SVA) [17] [19]:

$$\frac{\partial \bar{E}_2}{\partial t} + c\frac{\partial \bar{E}_2}{\partial x} = ig'\bar{\rho}_{20} \qquad (24)$$

g' is a propagation constant given by:

$$g' = \frac{2\pi}{\varepsilon_0}Ng_2(\omega_2 + \delta_2) \qquad (25)$$

Where ε_0 is the vacuum electric constant, N the atomic dipole density and c the light velocity of light. The condition for soliton propagation of the field amplitude is expressed as:

$$\bar{E}_2(x, t) = \bar{E}_2(x - v_gt) \qquad (26)$$

g' and $\bar{E}_2$ are assumed to be real .v_g can be identified with the group velocity of the solitons. We introduce a moving coordinate which propagates with the pulses' velocity.

$$z = x - v_gt \qquad (27)$$

and we have

$$\frac{\partial}{\partial t} = -v_g\frac{\partial}{\partial z}$$
$$\frac{\partial}{\partial x} = \frac{\partial}{\partial z} \qquad (28)$$

We assume that the two spontaneous emission rates are approximately equal: $\gamma_1 = \gamma_2 = \gamma$. We deal with the case of two photon resonance $\delta_1 = \delta_2$

The fact that $\bar{E}_2$ is real gives us:

$$\chi_{20} = 0 \qquad (29)$$

The complete set of the evolution equations for medium-fields interaction (Maxwell-Bloch equations) can be obtained from Maxwell equation and the system of evolution equations for the density matrix :

$$\frac{d}{dz}\chi_{10} = \frac{\delta}{v_g}\psi_{10} - \alpha_2\psi_{21} + \Gamma\chi_{10}$$

$$\frac{d}{dz}\psi_{10} = \frac{\delta}{v_g}\chi_{10} + \alpha_1(1 - 2\rho_{11} - \rho_{22}) - \alpha_2\chi_{21} + \Gamma\psi_{10}$$

$$0 = \frac{\delta}{v_g}\psi_{20} + \alpha_1\psi_{21}$$

$$\frac{d}{dz}\psi_{20} = \alpha_2(1 - \rho_{11} - 2\rho_{22}) - \alpha_1\chi_{21} + \Gamma\psi_{20} \qquad (30)$$

$$\frac{d}{dz}\chi_{21} = \alpha_1\psi_{20} + \alpha_2\psi_{10}$$

$$\frac{d}{dz}\psi_{21} = \alpha_2\chi_{10}$$

$$\frac{d}{dz}\rho_{11} = 2\,\alpha_1\psi_{10} - \Gamma(1 - \rho_{11} - \rho_{22})$$

$$\frac{d}{dz}\rho_{22} = 2\,\alpha_2\psi_{20} - \Gamma(1 - \rho_{11} - \rho_{22})$$

$$\frac{d}{dz}\alpha_2 = -\frac{d_2 g'}{v_g(c - v_g)}\psi_{20}$$

Where α_1 and α_2 are related to the field amplitudes by the following expressions:

$$\alpha_1 = \frac{d_1 \bar{E}_1}{v_g}$$

$$\alpha_2 = \frac{d_2 \bar{E}_2}{v_g} \tag{31}$$

Γ and g are defined as new constants :

$$\Gamma = \frac{\gamma}{v_g} \tag{32}$$

$$g = \frac{g' d_2}{(c - v_g)\, v_g} \tag{33}$$

The Maxwell-Bloch equations (30) are a system of differential equations. Our interest is in studying the evolution of the field α_2. The field 2 has a slowly varying amplitude, in this case, we can neglect the variation of the curvature and we can assume that the third and the forth order of derivation are negligible. Furthermore, if the driving field 1 is strong and the signal field 2 is weak, then the population of level 2 is almost equal to 1. After algebraic manipulations and differentiation of the Maxwell-Bloch equations (30) , we obtain a nonlinear second-order differential equation:

$$\frac{d^2 u}{dz^2}\frac{du}{dz}u^2 = k\frac{du}{dz}u^3 + \eta_3 u^3 + \eta_2 u^2 + F u^2 (\frac{du}{dz})^2 + \eta_1 u + Q(\frac{du}{dz})^2 + M \tag{34}$$

Where u is the dimensionless variable describing the field evolution of the field α_2 normed to the field α_1:

$$u = \frac{\alpha_2}{\alpha_1} \tag{35}$$

and k represents a normalized coupling constant:

$$k = \frac{d_2}{v_g}\frac{g'}{(c - v_g)} \tag{36}$$

The expressions of the constants η_i, Q and M are:

$$
\begin{aligned}
\eta_1 &= \frac{-2C_3}{\Gamma\alpha_1} \\
\eta_2 &= \frac{2kC_2}{\Gamma\alpha_1} \\
\eta_3 &= \frac{-kC_3}{\Gamma\alpha_1} - \frac{c}{4\alpha_1^2} \\
Q &= -\frac{\Gamma}{k} - \frac{4\alpha_1^2}{\Gamma k} \\
F &= \frac{\Gamma}{k} - \frac{2\alpha_1^2}{\Gamma} \\
M &= \frac{\delta C_1}{v_g \alpha_1^3} + \frac{\Gamma C_2}{\alpha_1^3} + \frac{4C_2}{\Gamma\alpha_1}
\end{aligned}
\tag{37}
$$

c is the light velocity and C_i represent the integration constants verifying:

$$
\begin{aligned}
C_1 &= \left(\frac{dJ_{21}}{dz}\right)_{z=0}\left(\frac{d\alpha_2}{dz}\right)_{z=0} \\
C_2 &= \left(\frac{dR_{21}}{dz}\right)_{z=0}\left(\frac{d\alpha_2}{dz}\right)_{z=0} \\
C_3 &= \left(\frac{dR_{11}}{dz}\right)_{z=0}\left(\frac{d\alpha_2}{dz}\right)_{z=0}
\end{aligned}
\tag{38}
$$

To solve the differential equation (34), we introduce a new variable $p(u)$ which is defined as:

$$
p(u) = \frac{d}{dz}u(z(u))
\tag{39}
$$

This give us a non-linear first-order differential equation:

$$
p^2\left(u^2\frac{dp}{du} - Fu^2 - Q\right) - ku^3 p = \eta_3 u^3 + \eta_2 u^2 + \eta_1 u + M
\tag{40}
$$

In the case of the strong stationary laser field applied on the Stokes transition we have:

$$
u^3 \ll 1
\tag{41}
$$

and we can consider $p(u) \approx A_0 + A_1 u + A_2 u^2$. The coefficients A_i are determined from the differential equation (40):

$$
\begin{aligned}
A_0 &= \frac{-C_2 k}{\alpha_1^3} \\
A_1 &= \frac{C_3 k}{2\alpha_1^3} \\
A_2 &= \frac{-C_2 C_3 k^3 \Gamma}{8\alpha_1^8}
\end{aligned}
\tag{42}
$$

Our aim is to find the evolution of the field, we must find a relation between u and z. From the definition of the p function we obtain the following relation between the normed field u and the local coordinate z:

$$
\begin{aligned}
z &= \int \frac{du}{p} \\
&= \int \frac{du}{A_0 + A_1 u + A_2 u^2} \\
&= \frac{-2}{\sqrt{\beta}} \tanh(\frac{2A_2 u + A_1}{\sqrt{\beta}})
\end{aligned}
\tag{43}
$$

Where β is given by:

$$
\beta = \frac{C_3^2 k^2}{4\alpha_1^6} - \frac{C_3 C_2^2 \Gamma k^4}{2\alpha_1^{11}}
\tag{44}
$$

The expression of the integration constants C_i are given by the following equations:

$$
\begin{aligned}
C_2 &= \left(\frac{dR_{11}}{dz}\right)_{z=0} \frac{\alpha_1^3}{2k} \\
C_3 &= (\frac{dR_{11}}{dz})_{z=0}^2 \frac{\alpha_1^3}{2k} \frac{1}{\left(\frac{dR_{21}}{dz}\right)_{z=0}}
\end{aligned}
\tag{45}
$$

Herewith we obtain the soliton pulse shape:

$$
u(z) = \frac{-A_1}{2A_2} + \frac{\sqrt{\beta}}{2A_2} \tanh(\frac{z\sqrt{\beta}}{-2})
\tag{46}
$$

4. Soliton Velocity

In order to calculate the velocity of the soliton, we determine, first the maximum amplitude of the soliton. The maximum amplitude of the soliton is given by the following relation:

$$
U_{\max} = \frac{k\Gamma}{\alpha_1^2} \frac{(\frac{dR_{11}}{dz})_{z=0}^5}{64} \frac{1}{(\frac{dR_{21}}{dz})_{z=0}^2}
\tag{47}
$$

From the equations that define k, Γ and α_1 (32) (33), we deduce the soliton velocity:

$$
v = c \left(1 - \Phi \frac{(\omega_2 + \delta)}{\bar{E}_1^2 U_{\max}}\right)
\tag{48}
$$

where Φ represents the speed deceleration factor:

$$
\Phi = \frac{\gamma \pi d_2 N g_2 (\frac{dR_{11}}{dz})_{z=0}^5}{32\varepsilon_0 d_1^2 (\frac{dR_{21}}{dz})_{z=0}^2}
\tag{49}
$$

The equation (48) allows us to conclude that the soliton velocity can be controlled by the amplitude of the stationary control field, the maximum amplitude of the soliton and by the detunings between the two atomic transitions and the laser sources.

The speed of the soliton decreases when :

-the detuning increases

-the maximum amplitude of the soliton decreases

-the amplitude of the laser field decreases.

The soliton velocity is obviously $\geqslant 0$, imposes an existence condition for the soliton:

$$\delta \leqslant \frac{\bar{E}_1^2 U_{\max}}{\Phi} - \omega_2 \tag{50}$$

For the special case and trickily chosen atomic parameters the soliton can be stopped for:

$$\delta = \frac{\bar{E}_1^2 U_{\max}}{\Phi} - \omega_2 \tag{51}$$

This case has potential applications in logic quantum gates and logic quantum memories.

Conclusion

In summary, we have presented the analytical expression of the soliton shapes that can propagate through an absorbing three-level atoms excited by a stationary laser field. These solitons propagate through the medium with a velocity depending mainly on the amplitude of the stationary control field, the maximum amplitude of the soliton and the detuning. Especially, we show that for a given value of the detuning, the soliton can be completely stopped.

These results are useful in optical data communication, where the optical fibre can be modelled as an absorbing three level system [20]. The advantage of soliton in supporting data information is the invariance of their shape which minimizes the noise effect, this is usually the origin of the signal defects. Besides, solitons propagate without dispersion. Therefore, we can send the optical information with high bit rate.The possibility to completely stop the soliton has potential application in the development of quantum memories.

Acknowledgment

We would like to thank Professor Alfons Stalhlhofen at the university of Koblenz for his fruitful advice and stimulating discussions.

References

[1] Remoissenet, M.; waves called solitons concepts and experiments. Springer-Verlag, 2003.

[2] Liu, J.; Li, R.; Xu, Z. *Physical Review A*. 2006, vol.74, 43801.

[3] Mancic, A.; Hadzievski, L.; Skoric, *MM. Physics of Plasmas*. 2006, vol.13, 52309.

[4] Kasamatsu, K.; Tsubota, *M. Phys. Rev. A.* 2006, vol 74, 013617.

[5] Brizhik, LS.; Eremko, *AA. Electromagnetic Biology and Medicine*. 2003, vol.22, no.1, 31-9.

[6] Van capel, P. J. S.; Muskens, O. L.; Husselink, E. W.; Dijkhuis, *J. I. Physica Status Solidi C*. 2004, vol 1, issue 11, 2749.

[7] Rickett, DS.; Ham, D. *IEEE International Solid-State Circuits Conference* 2006, 10. Piscataway, NJ, USA.

[8] Zong, F. D.; Dai, C. Q.; Yang, Q.; Zhang, J. F.; *Physica Sinica*. 2006, vol.55, no.8, 3805.

[9] Hau, L.V et al. *Nature*. 1999, 397, 594.

[10] Liu, C.; Dutton, Z.; Behroozi, C.H.; Hau, L.V. *Nature*. 2001, 409, 490.

[11] Sun, C.P.; Li, Y.; Liu, X. F. *Phys. Rev. Lett.* 2003, vol 91, 147903.

[12] Schmidt, H.; Imamoglu, *A. Opt. Lett.* 1996, vol 21, 1936.

[13] Harris, *S. E. Phys. Today.* 1997, 50, vol 7, 36.

[14] Boller, K. J.; Imamogluand, A.; Harris, *S.E. Phys. Rev. Lett.* 1991, vol 66, 2593.

[15] Alzetta, G.; Gozzini, A.; Moi, L.; Orriols, G. *Nuova Cimento.* 1976, B 36, 5.

[16] Gray, H.R.;.Whitley, R.Mand.; Stroud, *C.R. Optics Letters.* 1978, Vol 3, No 6, 218.

[17] Eberly, *J.H. Quantum Semiclass.* 1995, vol 7, 373.

[18] Mandel, L.; Wolf, E. *Optical Coherence and Quantum Optics,* Cambridge University Press, 1995.

[19] Allen, L.; Eberly, J.H.; *Optical Resonance and Two-Level Atoms.* Dover, New York, 1987.

[20] Guo, Y.; Kao, C.K.; Li, E.H.; Chiang,K.S.; *Nonlinear Photonics.* Series in Photonics, Springer 2002.

In: Perspectives in Optics Research
Editor: Jeffrey M. Ringer, pp. 31-49

ISBN: 978-1-61122-934-9
© 2011 Nova Science Publishers, Inc.

Chapter 3

IN-LINE HOLOGRAM RECONSTRUCTION BY USING ITERATIVE ALGORITHMS

*Y. Zhang**

Department of Physics, Capital Normal University,
Xisanhuan Beilu 105, 100037 Beijing, China

ABSTRACT

Digital holography is a quickly developing technology for high speed imaging. In-line holography has been extensively investigated due to it can effectively utilize the space-bandwidth product of digital recording instruments. However, since the reference wave and the object wave are overlapped during the hologram is recorded; the directly inverse Fresnel transform can not present the original object well. The reconstructed image is blurred by its conjugated image and the zero-order diffraction of the hologram. This shortcoming limits the application of the in-line holography. Therefore, an effective reconstruction algorithm is important for generalizing its application.

In this chapter, some approaches based on the phase retrieval algorithm for in-line hologram reconstruction are reported. Firstly, the Yang-Gu algorithm and the GS algorithm are used to restructure pure absorption and pure phase object from their in-line holograms, respectively. The differences between these two algorithms are analyzed. Then the GS algorithm is extended to reconstruct whole optical field from double or multiple holograms. At last, a new approach for reconstructing object from a hologram series is presented. Experimental results show that all these methods can reconstruct original object well.

Keywords: digital holography, in-line holography, reconstruction, iterative algorithm

1. INTRODUCTION

Holography is a powerful approach for storing and reconstructing both the amplitude and phase information of a wave front [1]. In the conventional approach the hologram is recorded

* E-mail address: yzhang@mail.cnu.edu.cn; Tel. and Fax: 86-10-68902178 on the photosensitive materials. The reconstruction is performed by illuminating the developed materials with the reference wave and the wet processing is necessary; therefore, the traditional holography is time consuming. Using electrical imaging instruments instead of the photosensitive materials, a new kind of technology called digital holography (DH) is proposed. In this technology, the hologram is digitalized and fed into a computer. The reconstruction is achieved in the computer. The DH has many advantages: It can provide both the amplitude and phase information of a wave front simultaneously. The reconstruction does not require the wet processing, therefore, the DH can get all information of an object in real time. Using a high-speed recorder, the DH can also record the dynamic varieties of an object, thus it can provide four-dimensional imaging. At present, the DH has already used in many fields such as deformation analysis [3], microscopy [4], particle measurement [5], object contouring [6], and so on.

However, the size of the object to be imaged is limited by the CCD's size and resolution. In-line holography uses one wave which severs as both the illuminating light and reference light simultaneously, therefore, it can effectively utilize the spatial band product of the CCD. Due to no beam splitter or mirror is required; this technology is quite suitable for short wavelength light imaging such as X-ray imaging. The directly inverse Fresnel transform of the in-line hologram will present the real image, the zero order term, and the conjugation image together. Since the reference light and the object light are overlapped during the hologram is recorded, the real image and its conjugation image can not be separated. Many researchers tried to solve this problem. Lai et. al. use off-axis reconstruction of the in-line hologram to avoid ghost image [7], but this method is limited to small object due to losing low frequency components. Phase-shifting method [8] can obtain the phase and amplitude information of a wave front by processing three or more holograms, however, the corresponding optical setup is quite complex and the phase shifter for recording each hologram must be controlled strictly.

In this chapter, some reconstruction methods based on the iterative phase retrieval algorithm are introduced. The YG algorithm [9] and the GS algorithm [10, 11] are used to reconstruct pure absorption object or pure phase object from their in-line holograms, respectively. The differences between these two algorithms in in-line hologram reconstruction are analyzed. Then the GS algorithm is extended to reconstruct whole optical field form double or multiple holograms. At last, a new approach for reconstructing object from a hologram series is presented.

2. FORMULAS USED IN RECONSTRUCTION

The in-line digital hologram recording system is schematically shown in Fig. 1. The object is illuminated by a plane wave. A part of light passes the object without any changes; other part suffers an intensity or phase modulation. Assuming that the distribution function of the object is $a(x', y')$, the intensity distribution recorded at the CCD surface will be:

$$I(x, y) = |A(x, y)|^2,$$

$$\text{(1)}$$

with

$$A(x,y) = \frac{1}{\lambda d}\exp(i\frac{2\pi}{\lambda}d)\int a(x',y')\exp\{\frac{i\pi}{\lambda d}[(x-x')^2+(y-y')^2]\}dx'dy',$$

(2)

where λ is the wavelength of the incident light and d is the distance between the object and the CCD surface. The reconstruction problem can be described as: Known the intensity distribution $I(x,y)$ of the hologram, how can one retrieval the distribution $a(x',y')$. Usually, there is no unique resolution for this problem. However, if some information about the object is known, for example, the object is a real or pure phase one; the reconstruction problem can be solved by using the phase retrieval algorithm.

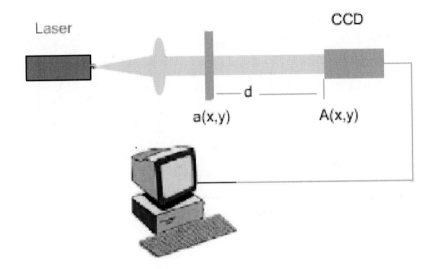

Figure 1. Digital hologram recording system.

Only one-dimensional case is considered here for the sake of brevity; all the formulas obtained here can be expanded to the two-dimensional case directly. In the matrix form, Eq. (2) can be rewritten as:

$$A(x) = \mathbf{G}\,a(x'),$$

(3)

where $\mathbf{G}$ expresses the Fresnel integral operator. If the system is a diffractive loss or nonparaxial one, $\mathbf{G}$ will be a nonuniatry matrix and $\mathbf{G}*\mathbf{G} = \mathbf{A} \neq \mathbf{I}$, where the superscript * represents the Hermitian conjugation and $\mathbf{I}$ is the identity transform. For reconstructing the object in a computer, the continuous wave functions must be sampled. Assume that the numbers of sampling pixels in the object and hologram planes are N_1 and N_2, respectively, then the distributions $a(x')$ and $A(x)$ can be expressed by vectors and the operator $\mathbf{G}$ corresponds to a matrix:

$$a_l = \rho_{1l}\exp(i\varphi_{1l}), \quad l=1, \ 2, \ 3, \ ..., \ N_1,$$

(4a)

$$A_m = \rho_{2m}\exp(i\varphi_{2m}),$$

(4b)

$$A_m = \sum_{l=1}^{N_1}\mathbf{G}_{ml}a_l \quad m=1, \ 2, \ 3, \ ..., \ N_2.$$

(4c)

where ρ_1, φ_1, ρ_2, and φ_2 are amplitude and phase distribution of the $a(x')$ and $A(x)$, respectively. In order to describe the closeness of $\mathbf{G}a(x')$ to $A(x)$, a distance measure is defined as

$$D(\rho_1,\varphi_1,\rho_2,\varphi_2) = \|A - \mathbf{G}a\| = \left[\sum_{m=1}^{N_2}|A_m - (\mathbf{G}a)_m|^2\right]^{1/2}.$$

(5)

If $D=0$, then $A(x)$ will exactly equal to $\mathbf{G}a(x')$ Thus the reconstruction problem can be formulated as the search for the extreme of the function D^2 with respect to the function arguments ρ_1, φ_1, ρ_2 and φ_2, i.e.

$$\begin{aligned}\Delta_{\rho_1}D^2 = 0, \quad &\Delta_{\varphi_1}D^2 = 0,\\ \Delta_{\rho_2}D^2 = 0, \quad &\Delta_{\varphi_2}D^2 = 0,\end{aligned}$$

(6)

where $\Delta_\xi D^2$ represents the functional variation of D^2 with respect to the function argument ξ [11].

It is easy to obtain the following equations by using matrix algebra

$$\rho_{1k} = \frac{1}{\mathbf{A}_{kk}}Abs\left[\sum_{j=1}^{N_2}\mathbf{G}_{jk}^*\rho_{2j}\exp(i\phi_{2j}) - \sum_{j\neq k}\mathbf{A}_{kj}\rho_{1j}\exp(i\phi_{1j})\right],$$

(7a)

$$\varphi_{1k} = Arg\left[\sum_{j=1}^{N_2}\mathbf{G}_{jk}^*\rho_{2j}\exp(i\phi_{2j}) - \sum_{j\neq k}\mathbf{A}_{kj}\rho_{1j}\exp(i\phi_{1j})\right],$$

(7b)

$$\rho_{2k} = Abs\left[\sum_{j=1}^{N_1}\mathbf{G}_{kj}\rho_{1j}\exp(i\varphi_{1j})\right],$$

(7c)

$$\varphi_{2k} = Arg\left[\sum_{j=1}^{N_1} \mathbf{G}_{kj}\rho_{1j} \exp(i\varphi_{1j})\right].$$

(7d)

If **G** is a unitary operator, Eqs. (7) will reduce to those of the GS algorithm

$$\rho_1 = Abs(\mathbf{G}*A),$$

(8a)

$$\varphi_1 = Arg(\mathbf{G}*A),$$

(8b)

$$\rho_2 = Abs(\mathbf{G}a),$$

(8c)

$$\varphi_2 = Arg(\mathbf{G}a).$$

(8d)

Generally, Eqs. (7) and (8) have no analytical solution and should be solved numerically. The corresponding algorithms are called YG iteration algorithm and GS iteration algorithm, respectively. Eqs. (7a) and (7d) are used for the YG algorithm and Eqs. (8a) and (8d) are used for the GS algorithm to reconstruct pure absorption object. Eqs. (7b) and (7d) are used for the YG algorithm and Eqs. (8b) and (8d) are used for the GS algorithm to reconstruct pure phase object. The accuracy of the solution is evaluated by the sum-squared error (SSE), defined as:

$$SSE = \frac{\sum\left[\rho_2 - \left\|\mathbf{G}\rho_1^n \exp(i\varphi_1)\right\|\right]^2}{\sum \rho_2^2},$$

(9)

with ρ_1^n denotes the nth iteration for ρ_1.

3. DIFFERENCES BETWEEN THE YG AND GS ALGORITHMS

In this section, some numerical simulations are shown to compare the YG algorithm and the GS algorithm for pure absorption object reconstruction. The model object is presented in Fig. 2. It is a pure absorption object. The parameters of the simulating system are selected as: The wavelength of the illuminating light is $\lambda = 0.532\,\mu m$, the sampling ranges in the object and hologram planes are $a_1 = a_2 = 6.7\times1024\,\mu m$, and the numbers of sampling pixels in the object and hologram planes are $N_1 = N_2 = 1024$. It is found that only for $d = 8.64\times10^4\,\mu m$ the corresponding parameters satisfy the relation $N_1\lambda d / a_1 a_2 = 1$, thus the operator **G** is unitary [12]. The operator **G** will become more nonunitary with distance d increasing.

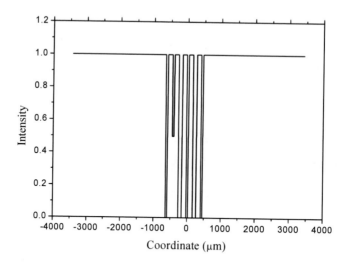

Figure 2. One dimensional object used in simulations.

In order to give a qualitative description, the ratio B/C is used to measure the degree of unitarity of the transform operator **G**. The quantity C and B is defined as the average value of the diagonal elements and of the off-diagonal elements of the matrix **A**

$$C = \frac{1}{N_1}\sum_{i=1}^{N_1}|A_{ii}| \tag{10}$$

$$B = \frac{1}{N_1(N_1-1)}\sum_{i=1}^{N_1}\sum_{j\neq i}^{N_1}|A_{ij}| \tag{11}$$

If the transform is unitary, one has $B=0$ and $C=1$. When **G** is nonunitary, one has $B\neq 0$, the lager the ratio B/C, the stronger the nonunitary of the transform. The variation of B/C with recording distance d is shown in Fig. 3. It can be seen that only when the recording distance is near to $8.64\times10^4\,\mu m$, the ratio B/C close to zero, thus the transfer is unitary. When d is lager than $8.64\times10^4\,\mu m$, the ratio B/C increases monotonically with d expect for the existence of a few links.

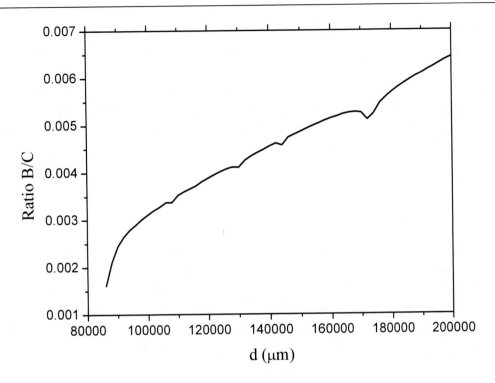

Figure 3. Variation of the ratio B/C with recording distance d.

Figure 4 describes the relationship between the SSE and the recording distance d for the YG algorithm and the GS algorithm, respectively. It can be found that the SSE for the GS algorithm increases quickly with d increasing and the corresponding reconstruction quality becomes worse. However, the SSE for the YG algorithm can keep a small value until the distance d is bigger than $1.7 \times 10^5 \mu m$. When the distance is selected as $d = 8.64 \times 10^4 \mu m$, the corresponding value of B/C is 9.48×10^{-5} and the system is nearly unitary, both the GS and YG algorithms can provide good reconstruction results. The retrieved results obtained are shown in Fig. (5a) by using the GS algorithm and in Fig. (5b) by using the YG algorithm. It can be found that the original object has been recovered quite well expect for small oscillations. The corresponding SSEs are 1.12×10^{-6} for the GS algorithm and 1.08×10^{-6} for the YG algorithm. This result are as expected due to two algorithms are identical when the system is unitary. For $d = 9.50 \times 10^4 \mu m$, the corresponding B/C is 2.85×10^{-3} and the system is nonunitary. The reconstructed results using two algorithms are shown in Fig. 6. The result obtained by using the GS algorithm in Fig. (6a) has stronger oscillations than that obtained by using the YG algorithm in Fig. (6b). The corresponding SSEs are 1.28×10^{-3} for the GS algorithm and 3.40×10^{-4} for the YG algorithm after 200 iterations.

Experiments have also been carried out to test these two algorithms [13]. The setup shown in Fig. 1 is used to record the holograms. The working wavelength is again $\lambda = 0.532 \mu m$, and the used CCD has 1300×1030 pixels with pixel size

$6.7\,\mu m \times 6.7\,\mu m$. Only 1024×1024 pixels are used for reconstruction. The used object is a cross made with two hairs. The hologram recorded at $d = 8.64 \times 10^4\,\mu m$ is shown in Fig. 7(a). Fig. 7(b) and 7(c) present the recovered images obtained by the GS algorithm and the YG algorithm after 5 iterations, respectively. Although the illuminating light is not so uniform and the hologram has been blurred by the disturbed specks, both two algorithms can recover the original object well. The corresponding SSEs are 7.52×10^{-4} for the GS algorithm and 7.51×10^{-4} for the YG algorithm. The reconstructed results obtained in a nonunitary system are also presented for a clear comparison. Fig. 8(a) shows the recorded hologram and Fig. 8(b) and 8(c) present the reconstructed object obtained by the GS algorithm and the YG algorithm after 5 iterations, respectively. It can be seen that Fig. 8(b) has stronger oscillation in the background. The corresponding SSEs are 8.96×10^{-3} and 6.21×10^{-3} for the YG algorithm and GS algorithm, respectively.

The reconstruction of a pure phase object has also been carried out experimentally [14]. The sample used is the logo of the *Institut für Technische Optik* shown in Fig. 9, it has a step height of $\sim 330nm$ at the location of underlying glass substrate, and the corresponding phase difference is $\sim 1.8rad$. The widths of the lines are $40 \sim 60\mu m$ depending on the difference parts in the logo. The distance between the object and CCD is $d = 8.64 \times 10^4\,\mu m$. Eqs. 8(b) and 8(d) are used to reconstruct the original object. Fig. 10 presents the reconstructed phase-contrast image obtained by the modified GS algorithm after500 iterations. The corresponding SSE is 0.1. This phase contrast reveals the topography of the sample. The height distribution is proportional to the reconstructed phase distribution. However, the values of recovered phase are limited in the internal of $\left[-\pi, \pi \right]$, the height difference greater than $\dfrac{\lambda}{2(n-1)}$ (n is the refractive index of the sample) will give rise to indeterminacy, but can be solved by used of phase-unwrapping methods.

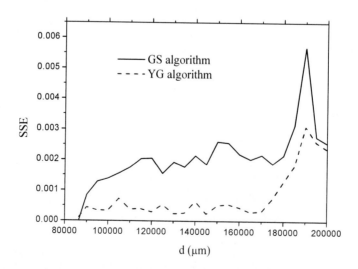

Figure 4. Relationship between the SSE and the distance d .

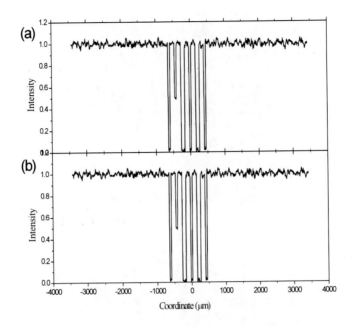

Figure 5. Reconstructed image obtained (a) by the GS algorithm and (b) by the YG algorithm after 200 iterations in an unitary system.

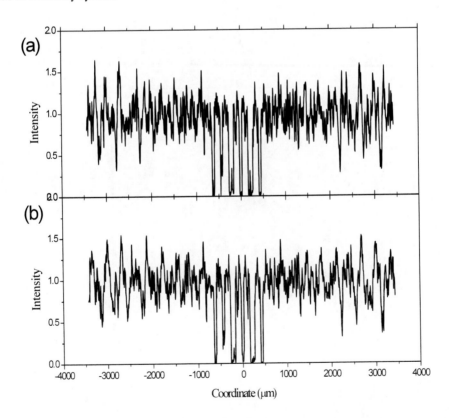

Figure 6. Reconstructed image obtained (a) by the GS algorithm and (b) by the YG algorithm after 200 iterations in a nonunitary system.

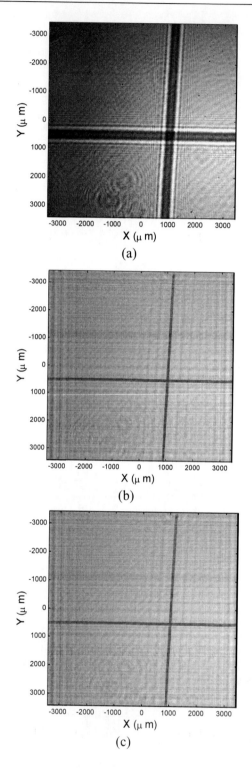

Figure 7. The reconstruction of real digital hologram recorded at $d = 8.64 \times 10^4 \, \mu m$. (a) Digitalized hologram, (b) result obtained by the GS algorithm after 5 iterations, and (c) result obtained by the YG algorithm after 5 iterations.

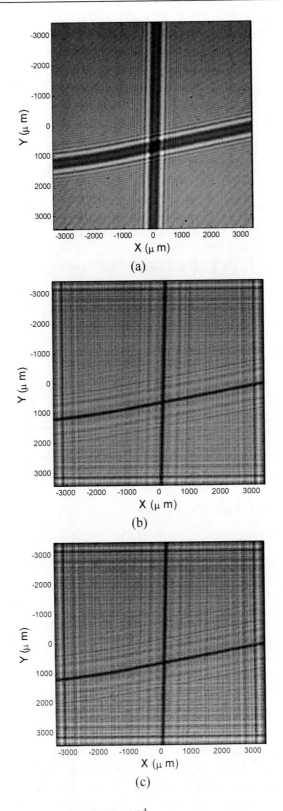

(a)

(b)

(c)

Figure 8. Same as Fig. 6 expect for $d = 9.50 \times 10^{4} \, \mu m$.

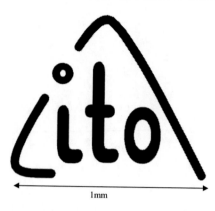

Figure 9. Pure phase object used in experiment.

Figure 10. Reconstructed pure phase object.

4. OBJECT RECONSTRUCTION FROM DOUBLE OR MULTI HOLOGRAMS

The object can also be reconstructed without know any prior information [15]. As shown in Fig. 11, two holograms H_0 and H_1 are recorded, then this two holograms are used to recover the phase distribution on the H_1 plane. Thus the recovered phase information and known amplitude distribution can be used to reconstruct original object. Furthermore, the whole wave field between the object plane and the CCD plane can also be obtained by using known wavefield. The object used in the experiment is also the logo of the *Institut für Technische Optik,* as shown in Fig. 9. Parts of three recorded holograms at distance $d = 7.40cm$, $8.15cm$, $8.90cm$ are shown in Fig. 12. The holograms (a) and (b) are used to retrieval the phase information. After 500 iterations, the SSE arrives 0.2956. The recovered amplitude and phase information at $d = 8.90cm$, i.e. the object plane, are shown in Fig. 13.

The variations of the amplitude and phase of the wave field with the propagation distance can also been achieved by reconstructing the complex wave field at difference locations. This character can be used to analyze the propagation of light in the phase-distorting media. However, it can be found from Fig. 13 that the recovered amplitude distribution at $d = 8.90cm$ is not an ideal plane wave, and the high frequency components have been filtered out. This is caused by the limitation of the size and resolution of the used CCD. Furthermore, the noise in holograms and errors in distance measurement also bring mistakes in the reconstruction. This algorithm is sensitive to the value of the initial phase, but this drawback can be overcome by using multi holograms reconstruction algorithm.

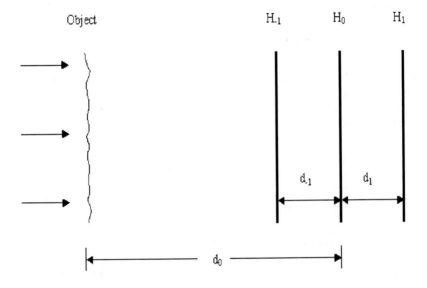

Figure 11. Schematic diagram of the iterative method for whole wave field reconstruction.

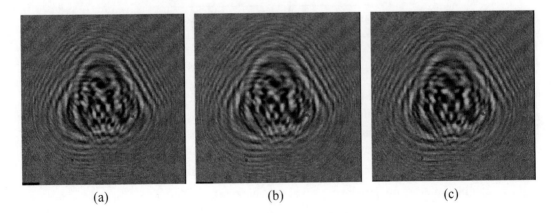

(a)   (b)   (c)

Figure 12. Holograms recorded at different distance. (a) $d = 7.40cm$ (b) $d = 8.15cm$ and (c) $d = 8.90cm$.

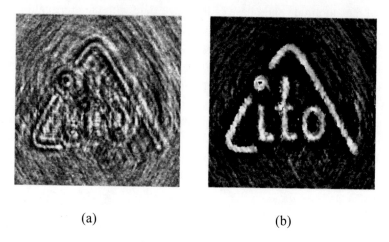

(a)   (b)

Figure 13. (a) Intensity and (b) phase distributions of the reconstructed wave field at $d = 8.90cm$, obtained by using double holograms.

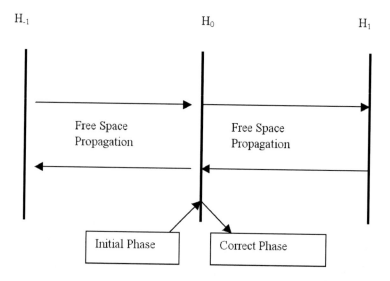

Figure 14. Schematic diagram of the iterative method of phase retrieval by using three holograms.

Since the algorithm based on two holograms is under constrained and susceptible to uniqueness concerns, more information are required to recover the phase information, thus more constraints are available and a deterministic solution is expected. As an example, three holograms are used to reconstruct the original object. As shown in Fig. 14, the iterative process begins from the central (H_0) plane, the amplitude obtained form H_0 and an arbitrary initial phase are employed to construct a wave function. This wave function propagates to the plane H_1, the phase distribution is kept unchanged and the amplitude is replaced by the amplitude information obtained from H_1, the new wave function propagates back to the plane H_0. Again, the amplitude is replaced by the amplitude information calculated from H_0 and the phase information is leaved unchanged. The new composed wave field continually propagates back to the plane H_{-1}, the amplitude is again corrected to the value obtained form H_{-1}. Then the corrected wave propagates to the central plane. The SSE is checked in every cycle and the iteration will be stopped if the SSE is sufficient small or the SSE no longer decreases.

(a)   (b)

Figure 15. (a) Intensity and (b) phase distributions of the reconstructed wave field at $d = 8.90cm$, obtained by using three holograms.

A demonstration reconstruction is preformed by using three holograms shown in Fig. 12. After 75 iterations, the SSE arrives at its minimal value of 15.96. Using the recovered phase distribution and the intensity distribution on the plane of H_1, the complex wave field at $d' = 8.90cm$ is reconstructed and presented in Fig. 15. Although three holograms are used, the SSE has not been reduced and reconstruction quality has not been improved comparing with the case by using double holograms. The noise in holograms and errors in distance measurements could be the main reasons. These errors or noise will be transferred into the recovered phase distribution in the iterative processes, thus the quality of the whole optical field reconstructed based on this phase distribution will be effected.

5. OBJECT RECONSTRUCTION FROM A HOLOGRAM SERIES

The original object can also be reconstructed from a series of holograms [16]. As shown in Fig.16, a series of holograms $H_n (n = 1, 2, 3, ..., N)$ are recorded. The distance between the first hologram and the object is d and the interval between two adjacent holograms is fixed as Δd. A propagation of the hologram H_1 is performed from d to $d + \Delta d$ for obtaining complex amplitude distribution of A_2, the phase distribution of A_2 is kept unchanged and the square root of the intensity distribution of hologram H_2 is used as the amplitude distribution for composing a new complex wave field A'_2. Then this new composed distribution A'_2 is performed a propagation from $d + \Delta d$ to $d + 2\Delta d$. The same procedure is repeated until all holograms are processed. Thus the wave field A_n can be obtained with high accuracy, and the original object can be reconstructed using the inverse propagation with a distance of $d+n\Delta d$. This approach is very similar to the algorithms for retrieving the phase information from two intensity distributions [10-14], in which the phase information is revised again and again by using the correct intensity distributions. This approach uses multiple holograms instead of two intensity distribution, and more information can be used to retrieve the phase information.

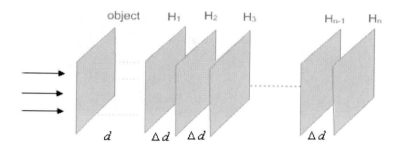

Figure 13. Schematic diagram for reconstruction from a hologram series.

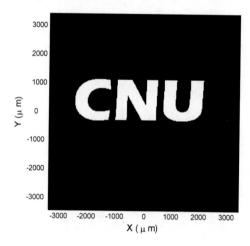

Figure 14. Original image used for simulation.

The object used in numerical simulations is shown in Fig. 14, its size is 6.8 x 6.8 mm². The parameters for the simulation are selected as follows: The wavelength of the illuminating light is also $\lambda = 0.532\,\mu m$; the area of the CCD window is $6.7\times1024\,\mu m \times 6.7\times1024\,\mu m$, the number of the pixels is 1024×1024, the distance from the object to the first hologram plane is $80\,mm$ and $\Delta d =100\,mm$. The relationship between values of c and n is shown in Fig. 15. It can be found that the value of c decreases with n increasing till $n = 27$, and then it increases with n. This phenomenon can be understood as follows: When the number of used holograms increase, more and more information is used to retrieve the phase of the wave field; therefore, the recovered wave fields will be more and more close to the real value. However, if the CCD is located too far from the object, it can not collect enough information due to the size limitation of the CCD, and thus the quality of the reconstructed wave filed will not be improved even more holograms are used.

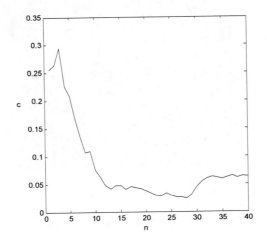

Figure 15. Relationship between the values c and n.

In our simulations, we used 27 holograms to reconstruct the original object; the reconstructed image is shown in Fig. 16. It can be see that the recovered object is nearly equal to the original one; all the details have been retrieved. For comparison purposes, the reconstructed image obtained by using the inverse Fresnel transform of the hologram recorded at distance $(80 + 27 \times 100)$ mm from the object is shown in Fig. 17, it is hard to recognize the original object due to the appearance of the ghost image. For a complex object, more holograms are required to reconstruct high quality image.

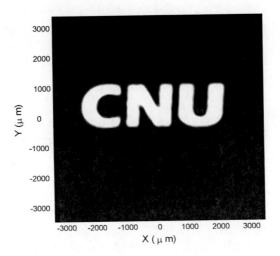

Figure 16. Image reconstructed by using our method.

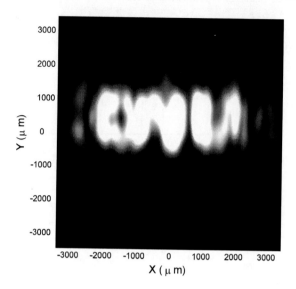

Figure 17. Image reconstructed by using the inverse Fresnel transform directly.

6. CONCLUSION

Several approaches based on iterative phase retrieval method are described in the presentation to reconstruct the full wave front form single, double, multi, and a series of in-line hologram(s). By processing these in-line holograms, the phase information in a hologram plane can be retrieved, thus the complete wave field can be achieved, and the original object can be reconstructed using the inverse propagation. Since that no reference light is required in these approach, the corresponding system is relatively simple. The signal-noise-ratio can be relatively improved. Both the computer simulations and experiment results demonstrate the validly of these approaches, the images reconstructed form experimentally recorded holograms exhibit good agreements with the corresponding original objects.

ACKNOWLEDGMENTS

The author would like to express his acknowledgements to Prof. H. J. Tiziani, Prof. W. Osten, and Dr. G. Pedrini for fruitful discussions and valuable suggestions.

REFERENCES

[1] D. Gabor, "A new microscopic principle," *Nature* (London) 161 777-778, 1948.
[2] J. W. Goodman and R. W. Lawrence, "Digital image formation form electronically detected holograms," *Appl. Phys. Lett.*, 11 77-79, 1967.

[3]   S. Schedin, G. Pedrini, H. Tiziani, A. K. Aggrarawl, and M. E. Gusev, "Highly sensitive pulse digital holography for built-in defect analysis with laser excitation," *Appl. Opt.*, 40 100-107, 2001.

[4]   U. Takaki and H. Ohzu, "Fast numerical reconstruction technique for high-resolutions hybrid holographic microscopy," *Appl. Opt.*, 38 2204-2211, 1999.

[5]   S. Murata, N. Yasuda, "Potential of digital holography in particle measurement," *Opics & Laser Techlonogy*, 32 pp. 567-574, 2000.

[6]   C. Wagner, W. Osten, S. Seebacher, "Directly shape measurement by digital wavefront reconstruction and multiwavelength contouring," *Opt. Eng.*, 39 pp. 79-85, 2000.

[7]   S. Lai, B. Kemper, G. V. Belly, "Off-axis reconstructions of in-line holograms for twin-image elimination," *Opt. Commun.* 169 37-43, 1999.

[8]   T. Zhang, I. Yamaguchi, "Three-dimensional microscopy with phase-shifting digital holography," *Opt. Lett.* 23 1221-1223, 1998.

[9]   R. W. Gerchberg and W. O. Saxton, "Practical algorithm for the determination of phase from image and diffraction plane pictures," *Optik* (Stuttgart), 35 237-246, 1972.

[10] J. R. Fienup, "Phase retrieval algorithms: a comparison," *Appl. Opt.*, 21 2758–2768, 1982.

[11] G. Z. Yang, B. Z. Dong, B. Y. Gu, J. Zhuang, and O. K. Ersoy, "Gerchberg–Saxton and Yang–Gu algorithms for phase retrieval in a nonunitary transform system: a comparison," *Appl. Opt.*, 33 209-218, 1994.

[12] Y. Zhang, B. Z. Dong, B. Y. Gu, and G. Z. Yang, "Beam shaping in the fractional Fourier domain," *J. Opt. Soc. Am. A* 15, 1114-1120 (1998).

[13] Y. Zhang, G. Pedrini, W. Osten, and H. J. Tiziani, "Image reconstruction for in-line holography using the Yang-Gu algorithm," *Appl. Opt.* 42 6452-6457, 2003.

[14] Yan Zhang, Giancarlo Pedrini, Wolfgang Osten, Hans J. Tiziani, "Phase retrieval microcopy for quantitative phase-contrast imaging", *Optik*, 115 94-97 (2004).

[15] Y. Zhang, G. Pedrini, W. Osten, and H. J. Tiziani, "Whole optical wave field reconstruction from double or multi in-line holograms by phase retrieval algorithm," *Opt. Express*, 11 3234-3241, 2003.

[16] De-Xiang Zheng, Yan Zhang, Jing-Ling Shen, Cun-Lin Zhang, "Wave field reconstruction from a hologram sequence", *Optics Commun.*, 249 7`3-77. (2005).

In: Perspectives in Optics Research
Editor: Jeffrey M. Ringer, pp. 51-82

ISBN: 978-1-61122-934-9
© 2011 Nova Science Publishers, Inc.

Chapter 4

PHOTOACOUSTIC SPECTROSCOPY OF NO2 AND SEVERAL APPLICATIONS

*Verónica Slezak**

CEILAP (CITEFA-CONICET)
Juan Bautista de La Salle 4397, 1603 Villa Martelli, Argentina

ABSTRACT

Photoacoustic spectroscopy is a very powerful tool for quantitative analysis of gaseous traces. Particularly, devices based on excitation by visible laser radiation allow sensitive detection of NO_2, which is a common urban pollutant. In this chapter, different photoacoustic systems are shown: some of them use resonant excitation of the acoustic modes of a cell by modulated laser radiation, other use Fourier analysis of the time response due to irradiation with pulsed lasers. The development of a simple and compact system for pollution control and its application to real measurements is much of interest. At the same time, very different issues get benefit from a vast experimental work concerning the photoacoustic detection of NO_2: the verification of theoretical models for photoacoustic phenomena, studies of surface phenomena (adsorption at the walls) and the determination of unknown absorption coefficients of certain molecules, which undergo multiphoton excitation, based on a precise determination of the setup calibration constant.

Pulsed photoacoustics is applied to the study of the Q-factor of a cylindrical acoustical cavity for determining the main kind of mechanism of the pressure wave's energy loss. A simple signal processing method for analysis in the frequency domain, which allows high resolution spectroscopy with low-cost equipment, is described.

Synchronous detection is used in photoacoustic setups based on the resonant scheme, which use either an amplitude modulated CW green laser or a high repetition frequency-doubled Q-switched Nd:YAG laser. The characteristics of different systems and their detection limit are shown. Further, the acquisition system is simplified through digitizing the pre-amplified signal from the microphone by the sound card of a PC, giving place to low-cost and easy-handling equipment, ideal for field measurements. Some of these systems are applied to the determination of the NO_2 content in car exhausts and the quality of catalytic converters. Moreover, a one-dimensional model for cell design and

---

* E-mail address: vslezak@citefa.gov.ar

optimization of S/N ratio is further verified by measurements on NO_2-air mixtures enclosed in a specially built acoustic cavity with a detached Helmholtz resonator.

1. INTRODUCTION

1.1. Optical Techniques for Gas Traces Detection

One of the big problems of our planet is the tremendous air pollution caused by human activity. Its control and prevention is of great importance for the international community, as shown, for example, by the Montreal protocol concerning substances that affect the ozone layer, such as halofluorocarbons and fluorocarbons. In particular, the pollution in urban areas mainly consists in gases produced by combustion of vehicles, like nitrogen oxides (NO_x), carbon monoxide and ozone. In this sense, the development of sensitive techniques for detecting traces of atmospheric gaseous compounds is very important.

A great number of methods have been applied with this purpose. Their requirements are diverse:

- High sensitivity
- Multicomponent capability, selectivity to different species
- Large dynamic range
- Real-time determinations
- Compactness, easy handling and mobility

If a great part of these characteristics were available at the same time as well as low-cost equipment, we would have reached a very attractive tool.

Nowadays there are well established gas-detection methods, like chemiluminescence, gas chromatography combined with mass spectrometry and chemical or electrochemical methods. Recently, spectroscopic techniques have been intensely applied in this field due to the progress in the development of new tunable radiation sources and detection schemes. Here, the measurement of the contaminant's concentration is based on phenomena of light scattering or absorption due to the radiation-matter interaction. Some examples of such devices are differential optical absorption spectroscopy (DOAS), Fourier-transform spectrometers and light detection and ranging systems (LIDAR). Detection schemes based on absorption measurement are of different types: long-path absorption [1], photoacoustic (PA) [2] and cavity ringdown [3].

In particular, this chapter concerns PA spectroscopy, which allows the precise determination of very little amounts of absorbed energy that, commonly, are not able to be measured by radiometers. Briefly, the method consists of measuring the microphone signal due to the pressure wave originated from the energy released to the surrounding by a molecular species which has undergone radiative excitation. The development of reliable compact solid lasers emitting in different ranges of the electromagnetic spectrum appears as an interesting addition to this classic technique [4]. Furthermore, its revival comes together with the progress of the detection equipment, electronics such as digital lock-in amplifiers and A/D cards, as well as the signal processing methods by means of desk computers.

This spectroscopic technique, we are concerned with, may be divided in two big branches: *resonant* and *pulsed* photoacoustics. In the first case, the acoustic signal arises from the amplified excitation of resonances of a closed or open cavity by amplitude or wavelength-modulated laser radiation. The repetition rate coincides with one of those frequencies. The signal is acquired and processed by a lock-in amplifier locked to the modulation frequency of the laser beam. This detection technique allows obtaining a high signal-to-noise ratio but the sensitivity of the trace gas detection may be limited by background signals because of periodic heating of the windows [5]. Furthermore, the resonance frequencies may shift due to changes in temperature, and the modulation frequency of the laser beam needs a tracking system in order to lock to the acoustic resonance if maximum signal is sought. This is particularly critical with high-Q systems (high gain).

On the other hand, the pulsed PA technique has the advantage of allowing choosing a time-acquisition window where the heating problems are no longer important [6]. The frequency tracking is overcome since conversion of the time-acquired signal to the frequency domain allows getting all the information contained in the resonance peak in a single-shot. In addition, pulsed PA enables the use of laser sources that extend to the UV region and solid-state-lasers. Both configurations have been widely applied to monitor different gas traces.

1.2. Nitrogen Dioxide Pollution

The troposphere contains a large amount of inorganic and organic gaseous compounds present in the form of traces. Many of them are released by natural processes, like methane; other, such as volatile organic compounds and nitrogen oxides, by man´s activities.

The NO_2 molecule plays a very important role in the photochemistry of the Earth´s atmosphere. In the troposphere, it acts as a source of ozone through the reactions:

$$NO_2 + h\upsilon \ (\ \lambda < 400 \ nm \ ) \rightarrow NO + O \qquad (1)$$

$$O + O_2 \rightarrow O_3 \qquad (2)$$

These reactions may be reversed by the rapid reaction of NO with ozone:

$$NO + O_3 \rightarrow NO_2 + O_2 \qquad (3)$$

Reactions (1) (2) and (3), with their dependence on the concentration of nitrogen oxides ([NO_2] and [NO]) and the light intensity, determine the local concentration of ozone.

Another harmful chemical transformation is provided by the photooxidation of NO_x emitted by cars, which produces the rain of nitric acid, so called acid rain.

The atmospheric pollution due to gases coming from car exhausts and electric power plants contains an important portion of nitrogen dioxide. Home heaters and gas stoves can also produce nitrogen dioxide inside homes. Generally, in the legislation about allowed pollution levels, these contaminants are indicated as NO_x (NO_2 and NO) but the first one is particularly toxic.

EPA's health-based national air quality standard for NO_2 is 0.053 ppm (measured as an annual arithmetic mean concentration). The excess of NO_2 amounts in air may bring about severe problems in the population, such as:

a- Health Effects: Short-term exposure may cause increased respiratory illness in young children and harm lung function in people with existing respiratory illnesses. Long-term exposure may lead to increased susceptibility to respiratory infection and may cause alterations in the lung. Nitrogen oxides also can be transformed in the atmosphere to ozone or fine particulate soot - which are both associated with serious adverse health effects.

b- Environmental Effects: Nitrogen oxides help form acid rain. In addition, this pollutant can cause a wide range of environmental damage, including visibility impairment and eutrophication - that is, explosive algae growth which can deplete oxygen in water bodies, such as the Chesapeake Bay.

We can often observe a brownish-yellow layer above large cities. It´s due to nitrogen dioxide, which is the most abundant species in the atmosphere with absorption in the visible range. Let´s look for the origin of this colour. Photochemical pollution, "smog", is formed by small particles that are formed from the reaction of pollutant hydrocarbons with ozone, in the presence of sunlight. These particles are Mie scatterers of sunlight, with diameters greater than 200 nm. Also, a large component of photochemical pollution is NO_2. This gas strongly absorbs scattered radiation at wavelengths shorter than 430 nm; hence it acts as a blue filter, removing the shorter wavelengths. The brown colour is a composite of primary colours; qualitatively, brown can be produced by removing the blue colour from white (brown = white − blue). By selectively filtering out the blue light, large amounts of NO_2 will give the remaining scattered light a brownish colour. The apparent colour will change as the angle between the sun, the haze layer, and the observer changes. Photochemical haze is most visible in the anti-solar direction (ie., looking away from the sun, in the direction of propagation of sunlight) because Mie scattering tends to scatter light in the forward direction, with relatively little back-scattering.

1.3. Spectroscopy of the NO_2 Molecule

This molecule presents a particular spectroscopic feature. Laboratory experiments on the NO_2 spectroscopy may be complicated by the presence of the dimer N_2O_4, through the reaction: $2NO_2 \leftrightarrow N_2O_4$, which shows an absorption cross section below 400 nm. At room temperature and low partial pressure, the equilibrium constant shows a clear tendency to NO_2 presence [7].

The excitation diagram of the NO_2 molecule first electronic levels is shown in figure 1. In the range from 370 to 460 nm the absorption takes place from ground state to the $^2B_1$ and $^2B_2$ levels; above 600 nm the transitions are only to $^2B_2$ [8]. A strong coupling between the high vibrational levels of the X^2A_1 ground state and $^2B_2$ or $^2B_1$ gives place to a long collision-free lifetime ($\sim$50 µs) [9]. The $^2B_1$ quenching rate by collisions with N_2 is 11.6 times the $^2B_2$ quenching rate [10] and the quenching constant deduced from fluorescence measurements at 550 nm is 44 Torr^{-1} times the spontaneous decay rate. Thus, after excitation with a short laser pulse, in the presence of some Torr of N_2, no radiation losses are present. Thus almost all the optically stored energy contributes to the sample heating and to the acoustic wave generation, regardless whether the excited level belongs to the $^2B_2$ or $^2B_1$ state.

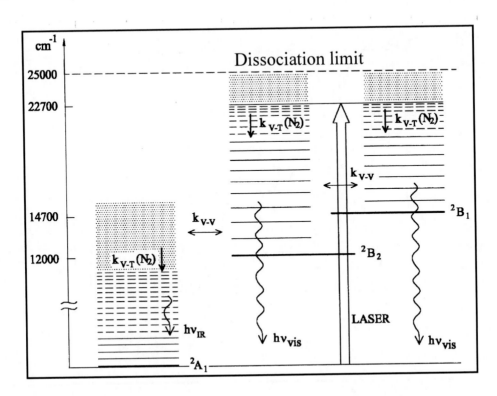

Figure 1. Excitation diagram of the NO_2 first electronic levels.

A thorough determination of the NO_2 absorption cross section in the spectral interval from 238 to 1000 nm at two temperatures and a resolution of 2 cm^{-1} can be found in Vandaele et al. [11]. The portion of the spectrum of interest through our experiments is shown in Fig. 2 [12]. It should be remarked that at $\lambda < 400$ nm dissociation of the NO_2 molecule becomes important [9].

1.4. Advantages of Visible PA Spectroscopy

In contrast to PA detection with infrared lasers, other gases present in the atmosphere do not interfere in this part of the spectrum. This fact gives place to a good selectivity.

Based on the resonant PA technique, very interesting results of the detection of trace gases, in particular NO_2, have been obtained with a CO laser [13]. The cell was specially designed to avoid the signal coming from the windows heating. The experimental set-up was dual-beam in order to correct the absorption measurements with respect to the presence of water vapor, which presents rather strong absorption bands between 5 and 6 μm. Formerly, other authors [14,15] have performed some short studies on the PA detection of NO_2 with other kind of visible laser sources. It is important to remark that the use of a laser source emitting in the visible rather than in the infrared range allows designing simple PA systems because the water absorption does not contribute to the signal and background signal due to windows heating is skipped.

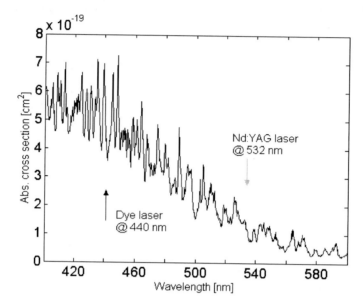

Figure 2. Absorption spectrum of NO_2; some visible-laser emission lines are indicated.

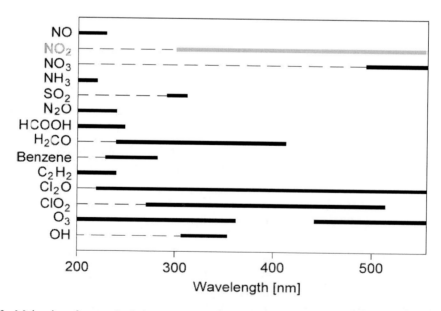

Figure 3. Molecules of atmospheric importance and their visible radiation absorption range.

Based upon spectral selectivity with respect to other species and the efficiency of the surrounding heating process, the development of PA-based devices for NO_2 monitoring seems to be very interesting. In this work, different experiments on resonant and pulsed PA detection of NO_2 traces in air, which were performed at the Center of Lasers and Applications

(CEILAP) in Buenos Aires, are described. For data processing, a method based on Fourier analysis of pulsed PA signals, which allows high resolution in the frequency domain, has been developed. Furthermore, some experiments, where samples taken from different pollution sources are analyzed, use a simple acquisition system for pulsed and resonant photoacoustics. It is based on recording the signal of the microphone, after amplification, through the sound port of a PC. With respect to acoustic cavity designs and calculations a numerical one-dimensional model, based on an equivalent electric-transmission-line, is applied to the design of a cell with a Helmholtz resonator associated to the main cavity. Its validity is checked with the experimental data obtained from NO_2 samples.

2. MATERIALS, COMPONENTS AND GAS HANDLING

For calibration of different PA systems, mixtures of NO_2 in chromatographic air at atmospheric pressure (NO_2, Liquid carbonic 99.5% further purified and air, L'Air Liquide 99.5%) or nitrogen (L'Air Liquide 99.999 %) were prepared at our lab. For this purpose before each filling cycle the resonators were evacuated down to 10^{-5} Torr through a greaseless vacuum system. The sample pressure was measured with a capacitance barometer (*MKS* mod. 122A, 1000 Torr). Very low concentrations were obtained through expansion to a known volume. Afterwards, each mixture was allowed stabilizing for several minutes before starting the data acquisition.

As NO_2 is a highly polar molecule, adsorption at the walls of the cell may affect all the measurements. This effect is strongly dependent on the wall's material, as former studies on H_2O or NH_3 have shown [16]. This matter will be further studied in this chapter. It imposes a limit on the time between collection of a sample and the measurement of the NO_2 concentration and suggests the convenience of working in flow regime for high precision determinations. In addition, most metals are highly reactive to this gas and should be avoided.

The microphones used in diverse PA experiments were cheap ear-aid electrets from Knowles Inc. They were adequate for NO_2 traces detection but suffered from corrosion when exposed during several hours to samples of some tens of Torr. They also seem to be mechanically fragile as regards sudden changes of pressure or in presence of large pressure gradients across the diaphragm as well as electrostatic charges.

3. PULSED PA EXPERIMENTS

3.1. Theory

For a cylinder of radius a and length L the wave equation of the acoustic pressure p, neglecting viscosity and thermal conduction effects, has a solution that is a superposition of eigenfunctions $p_j(r)$ of the form:

$$p_j(r) = J_m(k_r r)\cos(k_z z)\sin(m\varphi)$$

(4)

where J_m are the Bessel functions of order m, $k_r=X_{mn}/a$, $k_z=n_z/L$ and X_{mn} is the nth zero of the first derivative of J_m; $j=nmn_z$ is the subindex for the radial, azimuthal and longitudinal modes, respectively.

The frequency ω_j of the jth eigenmode is:

$$\frac{\omega_j}{2\pi} = v_{nmn_z} = \frac{s}{2}\left[\left(\frac{X_{mn}}{\pi a}\right)^2 + \left(\frac{n_z}{L}\right)^2\right]^{1/2} \qquad (5)$$

where s is the velocity of sound. For ideal gases $s = \sqrt{\dfrac{\gamma RT}{M}}$ where $\gamma = C_p/C_v = 1.4$ for diatomic molecules, T the temperature, R the gas constant and M the molecular weight.

The energy of the pressure wave decreases due to losses by thermal conductivity and viscosity in the bulk volume and near the walls. These effects can be taken into account by a complex ω_j:

$$\omega_j = f_j + i g_j \qquad (6)$$

where f_j is the resonance frequency of the mode v_{nmn_z}, g_j, g_j the damping rate of the sound wave and $2g_j$ the full width at half maximum of the resonance. The quality factor of the resonance is defined by $Q_j=f_j/2g_j$.

If the V-T relaxation rate (k_{VT}) is higher than the eigenfrequencies ω_j, the heating pulse after the laser excitation may be considered instantaneous compared to the time of the standing waves formation. Along typical experiments the mixtures with N_2 are such that k_{VT} is higher than 2.10^7 s^{-1} [9] and the value of f_j for the main detected frequency is around 7.10^3 s^{-1}. In this case, and if additionally k_{VT} is smaller than the inverse of the laser pulse duration, the Fourier transform of the total acoustic pressure gives a spectrum in the frequency domain, which is the sum over real and imaginary parts of Lorentzian profiles with their maxima at the resonator eigenfrequencies: [6,17]

$$p(r,\omega) \propto \sum_j p_j(r)\frac{S_j}{\omega_j(\omega-\omega_j)} \qquad (7)$$

where $S_j \propto \displaystyle\int_{V_{cell}} H(r)p_j^* dV \qquad (8)$

$H(r)$ is the spatial distribution of the heating pulse after the laser excitation and V_{cell} the volume of the cell. The heating pulse depends on the laser energy and the absorption cross section.

The equations (7) and (8) indicate that the amplitude of the PA power spectrum is conditioned by the exciting beam intensity and location, the sample absorption coefficient and the cell geometry. For the particular case of a perfect cylindrical geometry, the calculation of S_j is performed in ref. 18.

Using (6) and supposing $Q_j = f_j / 2g_j >> 1$, at a resonance, (7) becomes:

$$|p(r,f_j)| \propto \frac{|p_j(r)|\, S_j\, Q_j}{f_j^2} \tag{9}$$

The physical parameter g_j, which is included in Q_j, is the rate of volumetric and surface losses due to the sample viscosity (η) and thermal conductivity (λ) and has been analyzed by several authors [6,20].

The contribution of classical volumetric losses to Q^{-1} corresponding to the resonance (002) is [5]:

$$Q_{vol}^{-1} = \frac{4\pi}{\gamma P}\left[\frac{4}{3}\eta + (\gamma - 1)\frac{\lambda M}{C_p}\right] v_{002} \tag{10}$$

where P is the sample pressure, $\gamma = C_p/C_V$, η the viscosity, λ the thermal conductivity and M the molecular mass.

The thermal dissipation at the walls, due to non-adiabatic gas expansion and contraction, and viscous dissipation by the boundary conditions, for the second longitudinal mode, gives place to the expression [19]:

$$Q_{surf}^{-1} = \frac{1}{a}\left(\frac{RT}{M}\right)^{\frac{1}{2}}\left[\left(\frac{\eta}{P\pi v_{002}}\right)^{\frac{1}{2}} + (\gamma - 1)\left(\frac{\lambda M}{PC_p\pi v_{002}}\right)^{\frac{1}{2}}\left(1 + \frac{2a}{L}\right)\right] \tag{11}$$

The dependence of Q_j^{-1} on the sample pressure is given by the sum of the contributions of (10) and (11).

In the next section, based on these theoretical results, for a given irradiation condition, i.e. for a fixed S_j, we study experimentally the dependence of Q_j on the sample pressure. As (9) indicates, the spectrum amplitude at a resonance frequency f_j increases with Q_j. Thus, the knowledge of the main phenomena that determine the quality factor is very important for the optimization of a pulsed PA system which objective is a precise NO2 traces measurement.

3.2. Data Processing: High Resolution FFT [21]

As we stated before, a precise determination of the Q-factor is important. When the PA signal is analysed after transformation by the FFT algorithm, the result is a spectrum with a discrete frequency spacing that may be inadequate for a precise definition of the amplitude and width of the resonance peaks. The frequency resolution of the power spectrum should be approximately one tenth of the FWHM (Δf) of the resonance peak. Δf is related to the decay time (τ) of the PA signal due to losses near the walls of the resonator through the expression $\Delta f=1/\pi\tau$ [17]. Thus the signal should be acquired within a time window that takes the value $T=N/f_s=10/\Delta f=10\pi\tau$ where N is the total samples number and f_s the sampling frequency. This value of T is not practically achievable because of the noise-limited dynamic range of low signals that arise from traces detection. In our experiment the width of the resonance in the power spectrum for samples at atmospheric pressure takes a value of 10-15 Hz, corresponding to $\tau\sim$30-50 msec. Therefore, a spectral resolution of 1 Hz is required to determine the resonance peak amplitude. For very low concentrations, inside a window $T=7\tau$ the S/N ratio reaches the value 1 and the frequency resolution of the FFT spectrum is 3-5 Hz; it is clear that this frequency spacing will not precisely define the resonance.

Based on these considerations, in order to obtain improved resolution with respect to the direct application of the FFT to the whole signal in a window T, we developed an iterative processing method. Usually, the calculation of the FFT is numerically performed by means of the base-2 Cooley-Tukey algorithm ($N = 2^{\gamma}$, γ integer), which results in an important saving in computational time. When the sequence length is not a power of two, other FFT algorithms also allow significant time savings as long as N is highly composite, i.e. $N = r_1 r_2...r_m$ where r_i is an integer [22].

In each recurrence the appropriate FFT routine is applied to each of k different sets of samples of length N_k, which consist of the first successive $(N - j\delta N)$ samples, where δN is a very low number of samples ($\delta N<<N$) and $j = 0...k$-1. δN and k must be chosen according to the desired final frequency resolution and precision in the amplitude determination. Each iteration calculates the amplitudes for a frequency scale with spacing $(N_k\Delta)^{-1}$. The values of amplitude obtained for all the sets N_k and the corresponding frequencies give place to a spectrum with much smaller spacing than $1/N\Delta$. In practical cases, for acquisition cards of several thousand of samples, the variation of the calculated amplitude due to a different sample size is not significant because very good resolution is attained even if the window for the successive FFT is scarcely reduced, as will be shown below.

Many physical phenomena behave like damped oscillations in the time domain (radiation emission from an atom, acoustical waves in a medium with losses, etc.). Usually, the resulting analysis in the frequency domain is important. This will show a Lorentzian profile around the central frequency, whose width is related to the exponential decay time. Let us consider the following example:

$$f(t) = \exp(-at)\sin(2\pi f_0 t)$$

(12)

with a=50 Hz and f_0=1035 Hz. We choose f_s= 20 kHz and N=2000, so the window is five times the decay time.

Figure 4 shows the results obtained applying FFT over 2000 samples (open triangles) and the improved values derived by the high resolution FFT (HRFFT) method (dots) for N_k varying from 1960 to 2000 in δN steps of two samples. The amplitude of the Fourier transform of a truncated damped oscillation within a window T is reduced with respect to the Fourier integral by [1-exp($-aT$)]. The N_k variation from 1960 to 2000 allows the calculation of a corresponding amplitude change of 0.1%, which is very low and may be neglected through an experiment. For clarity, in figure 4 the calculated Lorentzian profile, which results from the Fourier integral of the function of equation (12), is not shown because it may not be resolved from the line that connects the dots, due to a near perfect coincidence.

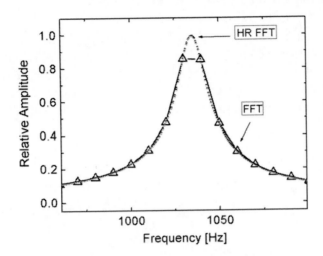

Figure 4. Comparison between the results from FFT and HRFFT application to a damped oscillation.

Let's look at an experimental application. A mixture of 5 ppmV of NO_2 in pure air at atmospheric pressure enclosed in a 30 cm long cylindrical resonator was excited by pulsed radiation at 532 nm from a Nd:YAG laser. The PA signal corresponding to the second longitudinal acoustic mode was acquired with a digital oscilloscope set at f_s=25 kHz and N=5000. A part of the high resolution amplitude spectrum, where the (002) resonance appears, obtained by application of the HRFFT to the signal (starting from N_k= 4900 up to N_k=5000 in steps of ten samples) is shown in Figure 5 (open circle). At the same time the spectrum that results from the direct application of the FFT to the N samples is shown (triangle). The best fit to a Lorentzian profile (straight line) is also shown. As one can see, the application of HRFFT brings about a precise amplitude determination without making use of the fitting with theoretically predicted functions.

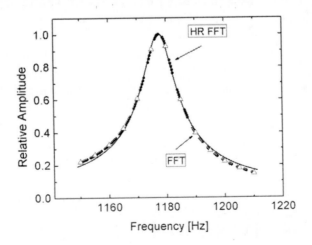

Figure 5. Comparison between FFT and HRFFT applied to an experimental PA signal.

3.3. Experimental

3.3.1. Excitation by a Tunable Dye Laser (430-470 nm)

The first PA studies at our lab were performed with excitation of NO_2/N_2 samples by a Nitrogen-pumped tunable dye laser [23]. The pulse energy was about 80 µJ; the laser linewidth was lower than 0.05 nm, the pulse length 8 nsec and the beam divergence 6 mrad. The acoustic resonator was a closed cylindrical cavity with a microphone (Knowles BT1759, sensitivity 10 mV/Pa at 1kHz) placed near its middle point. The cell internal diameter was 2.8 cm in front of the microphone and its average length 30 cm. The setup of this experiment is shown in Figure 6.

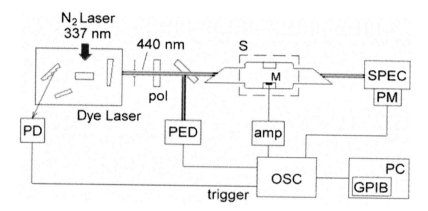

Figure 6. pol: polarizer, S: electrical shielding; SPEC: spectrometer; PM: photomultiplier; PD: photodiode for triggering; PED: pyroelectric detector.

The pulse energy was measured with a pyroelectric detector (Coherent mod. LMP5 with amplifier) for normalization and averaged over the adequate number of pulses (~60) to improve the signal-to-noise ratio. The acquisition system consisted of a Tektronix Mod.TDS540A digital oscilloscope connected through a GPIB interface to a computer for data processing. The oscilloscope was triggered with the laser pulse and the acoustic cavity and electrical connections were shielded so that the laser discharge slightly perturbed the beginning of the acoustic signal. In addition, after the signal acquisition, the laser beam was blocked and the average noise induced by the laser shot was registered. The analysis of the signal was performed on the difference between both registrations (averaged PA signal and noise).

The signal from the microphone was amplified and registered, carefully selecting the acquisition conditions. The sampling frequency and the total number of samples were set so that, with a square window, minimum deformations in the spectrum obtained by Fast Fourier Transform (FFT) were obtained. From the width of the fitted lorentzian function, for a mixture of 1 Torr NO_2 and 400 Torr N_2, we deduced a decay time of ~20 msec. This determined an acquisition window of at least 140 msec in order to obtain a pure Lorentz profile in the frequencies domain. Thus, for a sampling frequency of 25 kHz, fixed by the microphone's and amplifier's responsivity, that began to fall at frequencies higher than 10 kHz, a number of 5000 samples were chosen.

A typical signal, acquired according to the conditions cited above, is shown in Figure 7a-when a 1 Torr NO_2/400 Torr N_2 sample is irradiated with 80 μJ at 440 nm. For a good graphical resolution, only the first 100 ms are represented.

The corresponding Fourier spectrum, obtained with the HRFFT method, in the range including the main resonance, appears in Figure 7b. In the low-frequency range the most important resonance at around 1200 Hz corresponds, as expected, to the (002) mode, because the microphone is near the center of the cell z-axis, where the first longitudinal mode has a node. Other peaks possibly account for the cell irregularities, such as the constrictions at the center and at the ends. The fraction of spectral power distributed in these peaks for a given sample is constant as long as the beam radius and its position with respect to the center do not change.

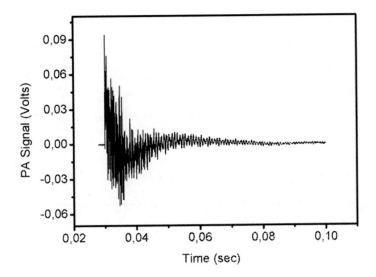

Figure 7a. Digitally acquired PA signal from a NO_2/N_2 mixture.

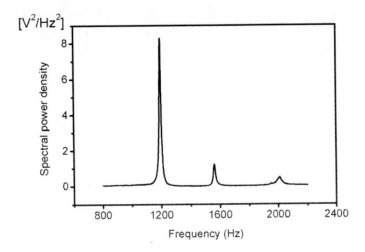

Figure 7b. Fourier spectrum of the PA signal.

The Q-factor of the acoustic cavity for different NO_2/N_2 mixtures was studied in order to determine the main processes of energy loss in the cell and to optimize the system for traces detection. The Q determination requires a precise measurement of the resonance peak's width and frequency. We looked for the best approximation of the (002) resonance by a lorentzian profile, as the theory predicts. The good quality of the fitting is evident in Figure 8, where the proposed function is shown superposed on the (002) mode power spectrum: this confirms the absence of perturbation by the neighboring peaks.

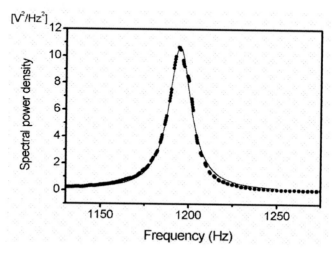

Figure 8. HRFFT of the PA signal (dots) and Lorentzian fitting (line).

The resonance frequency may shift due to thermal instabilities since the sound velocity depends on temperature. We verified that, with all the other experimental conditions fixed, the resonance peak's frequency increases with the room temperature at a rate of about 2 Hz/°C, therefore it is considered constant during the signal averaging, which may last at most a couple of minutes. From the fitting of the power spectrum around the (002) mode resonance frequency with a lorentzian profile we measured the width (Δv) and peak frequency (v) for various samples, with a fixed NO_2 pressure and different N_2 pressure values. The $1/Q_{002}$ factor as a function of pressure p was deduced from the ratio $\Delta v/v$ and plotted on Figure 9. At the same time we calculated Q^{-1} from the expressions (10) and (11). As the mixtures are in a ratio $NO_2{:}N_2$ lower than 1:10, we introduce in these expressions the thermal conductivity and viscosity that correspond to pure N_2, $\lambda = 6.10^{-5}$ cal cm^{-1} seg^{-1} K^{-1} and $\eta = 1.78 \ 10^{-4}$ g cm^{-1} seg^{-1}. For the conditions of our experiment calculation gives $Q_{vol}^{-1} \sim 10^{-4}/p$ and $Q_{surf}^{-1} \sim 0.19/p^{1/2}$ with p in Torr; thus, only surface losses are important for the range of pressures of this work. The theoretical results for Q_{surf}^{-1} vs pressure are also shown on Figure 9 with full line. There is a good agreement between the theory and experiment; the small undervalue of Q^{-1} at high pressure (<5%) is probably due to losses at the microphone or to the irregularities of the cell.

The results allow concluding that the presence of large N_2 amounts is very important to efficiently operate our cell for NO_2 traces PA detection. This point suggests that it should also be taken into account when the sample is picked up from a polluted atmosphere. Furthermore, based on these results and the expressions (9) and (11), we also conclude that the PA signal from NO_2 may be further improved by choosing a buffer gas with adequate thermodynamic (η and λ) and physical properties (M and k_{VT}).

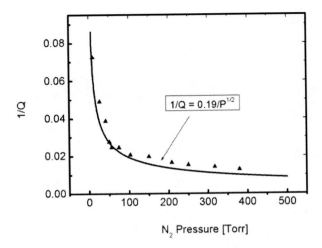

Figure 9- $1/Q_{surf}$ vs. pressure of N_2 for a fixed pressure of NO_2. Straight line: theoretical result, triangles: experimental data obtained as the ratio $\Delta\upsilon/\upsilon$ for the (002) resonance peak irradiating at 441 nm.

3.3.2. Excitation by the Second Harmonic of a Nd:YAG Laser (532 nm)

In the latter case, the average PA signal coming from the cell filled with pure N_2 takes the same value as when the laser radiation is blocked: there is no windows heating. Thus, the S/N ratio may be improved increasing the laser pulses energy. Therefore, in order to enhance the sensitivity of our PA system, a frequency-doubled electro-optically Q-switched Nd:YAG laser was tried [24], which delivered a pulse energy of 100 mJ at 10 Hz. As we look for linearity of our PA system, we have to prove if absorption saturation takes place in this case. Taking in account that the NO_2 absorption cross section at 532 nm is $\sigma = 1.5 \times 10^{-23}$ m^2 and the vibration-translation relaxation time of NO_2 in air at atmospheric pressure $\tau_{VT} = 1.9$ ns (see section 1.3), from a two-level-model we can calculate the saturation intensity from:

$$I_s = \frac{1}{2\sigma(v)\tau_{VT}}, \quad \tau_p \gg \tau_{VT} \tag{13}$$

where τ_p is the pulse duration. The result is $I_s = 6.4 \times 10^8$ W.cm^{-2}. In our case τ_p is 5 nsec and the beam diameter 6 mm, so the maximum available intensity without focusing is 6×10^7 W. cm^{-2}. However, as the structure of the electro-vibronic levels of this molecule is very complex, the validity of the two-level model for calculation of the saturation intensity is doubtful. Therefore, measurements of radiation absorption for a given mixture NO_2-N_2 and a wide laser energy range were performed; a very good linear correlation between incident and transmitted laser energy up to 100 mJ was found.

We based ourselves on these results to carry out PA measurements with this laser on a cylindrical cavity with particular dimensions that allow observation of radial resonances (length 10 cm, diameter 7.5 cm). An electret microphone (Knowles EM 4447) is glued on the inner wall of the latter cell at a node of the second longitudinal mode. The signal from the microphone is amplified, further digitized and averaged by an oscilloscope HP54616B at the same time as the signal coming from a pyroelectric detector which measures the laser pulse energy for normalization purposes. The signals are transferred to a computer for processing through the method of HRFFT and further analyzed. The spectrum shows the most intense peaks at 5700 and 10450 Hz, coincident with the theoretical resonances of the first and second radial mode respectively. Some weaker peaks correspond to combination modes. The experimental quality factor is 380.

The calibration of this system is performed for 60 mJ with different mixtures of NO_2-N_2, between 3 and 700 ppmV, as shown in Figure 10, where the amplitude of the peak at 5700 Hz minus the average corresponding to pure N_2 is plotted. A very good linear correlation (0.99) is obtained The red point corresponds to three times the standard deviation of the signal obtained for pure nitrogen and sets a detection limit of 0.02 ppmV

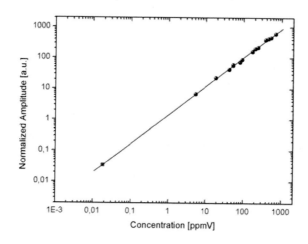

Figure 10. Calibration of the system.

This system was applied to the measurement of NO_2 traces concentration in air samples taken from our campus, directly into the cell through a desiccant ($CaSO_4$) and filters for 0.22 μm particles. The average over 3000 pulses is shown in Figure 11 together with the background of pure nitrogen. The deduced concentration value agrees with typical measurements carried out in the same season by other laboratories and published in official statistics of the town of Buenos Aires.

A problem which was detected during our measurements was the change of the PA signal during long signal acquisitions: this was due to adsorption/desorption of molecules on/off the walls. The NO_2 molecule is polar, therefore, like ammonia and water, presents this characteristic. The walls of the acoustical cavity may be considered a collection of sites where the molecules adhere particularly during the calibration process, and then, they act as a reservoir of molecules which interfere when real samples from urban areas, with very low concentrations, are measured. This fact introduces errors in the measurements and may be avoided or lowered by changing the material of the cell or adding gas-flow to the system [25]. The study of the rate of this process is useful for controlling the error of the measurement on steady samples and, even better, for skipping the problem by means of use of flowing samples.

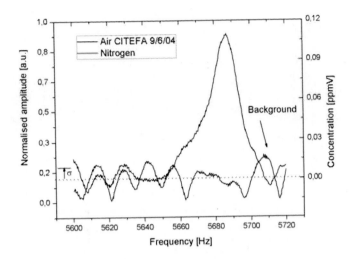

Figure 11. PA signal of a polluted air sample and background of pure N_2.

A qualitative theoretical model [26], which describes the adsorption-desorption dynamic equilibrium for a monomolecular layer in a cylindrical tube of radius r, is based on the law of mass-conservation:

$$\frac{\partial c_v}{\partial t} = -\frac{2}{r}\frac{\partial c_s}{\partial t} \qquad (14)$$

where c_v is the NO2 concentration [cm^{-3}], c_s [cm^{-2}] is the density of sites occupied by NO2 molecules. Furthermore, the dynamic exchange of molecules between surface and volume may be described as:

$$\frac{\partial c_s}{\partial t} = k_1 c_v \left(c_{s0} - c_s\right) - k_2 c_s \qquad (15)$$

where $c_{s0} - c_s$ is the density of available sites on the surface, $k_1 c_v$ is the adsorption rate and k_2 the desorption rate.

After the calibration process, the cell is cleaned by means of a high vacuum pipe and the density of occupied sites may be considered low ($c_s << c_{s0}$), so the equations system given by (14) and (15) may be approximated and has a simple solution. In this case the concentration of the gas, which detaches from the walls, shows a time dependence given by:

$$c_v\left(t\right) = \left(\frac{2 k_2 \alpha}{r} c_s\left(0\right)\right)\left(1 - \exp\left(-\frac{t}{\alpha}\right)\right) \qquad (16)$$

where the term in the first parenthesis represents $c_v\left(\infty\right)$ and $1/\alpha = 2/r\left(k_1 c_{s0} + k_2 r/2\right)$.

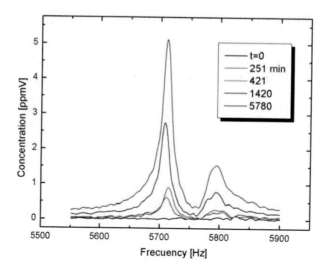

Figure 12. Evolution of desorption of a sample of pure N_2 at atmospheric pressure.

In Figure 12, the results of PA signal measurements are shown for a sample of pure Nitrogen enclosed in the cell, after a long lasting vacuum-cleaning process of the formerly calibrated cell.

The maxima of the (100) mode are indicated on plot of Figure 13 for different times and were fitted by the function (16), giving $c_v(\infty) \approx 5\,\mathrm{ppmV}$ and $\alpha \approx 2250$ min.

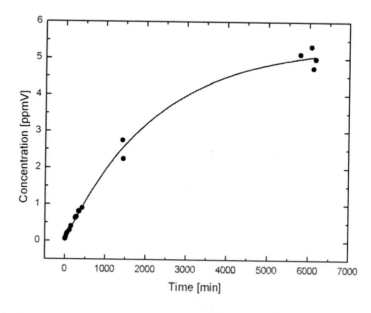

Figure 13. Evolution of NO_2 concentration due to desorption.

Furthermore, studies on the time evolution of prepared NO_2-N_2 samples lead to the conclusion that a flow regime of at least 1L Atm/h is required; else, enclosed samples should be measured in 20 minutes after collection, for obtaining an error lower than 5%.

4. RESONANT PA EXPERIMENTS

Acoustic pressure waves may also be generated by modulated laser radiation instead of laser pulses. If the modulation frequency coincides with resonance frequencies of the cell, amplification is obtained on that particular acoustic mode. Since the absorption cross section of NO_2 at 532 nm is 1.5×10^{-23} m^2, the samples used along our experiment (under 0.06 kPa) may be considered optically thin. In this case, the solution of the inhomogeneous wave equation for a perfect cylinder, where the heat source is the absorbed laser energy, leads to the expression for the amplitude of the n-th resonance at the microphone [27]:

$$A_n \cong K_n \frac{Q_n}{\omega_n} \alpha W \qquad (17)$$

where K_n contains the cell dimensions, the C_p/C_v ratio and the normalized overlap integral that describes the effect of the spatial overlap between the propagating laser beam and the pressure distribution of the n-th eigenmode of the resonator; α is the absorption coefficient for a fixed pressure, W the laser power density, ω_n the resonance frequency and Q_n the cell Q factor.

4.1. Simple and Cheap Acquisition System

Commonly, the signal from the microphone is synchronously processed by a lock-in amplifier. This way, a good sensitivity and high signal-to-noise ratio is achieved when light sources like CW CO, CO_2 and solid lasers are used. However, synchronous detection results in expensive equipment and lack of compactness. A simple, low-cost and compact signal acquisition and processing system for resonant PA spectroscopy which avoids the use of lock-in amplifiers is based on a desk or laptop computer with standard audio input. The performance of the system designed at our lab and its limitations are studied for PA detection of NO_2 in air at atmospheric pressure with a chopped multiline Argon laser and compared to the conventional lock-in-based-system [28].

The PA signal from the microphone for a given NO_2-N_2 mixture, measured at an arbitrary chopper frequency f_m, is amplified before entering into a PC through the sound port and digitally converted by the sound card (Sound Blaster PCI 128, full-scale sensitivity 200 mV) with a resolution of 16 bits at a sampling rate of 44.1 kHz. The signal is acquired during a time T and analyzed in the frequency domain after transformation by the Fast Fourier Transform (FFT). The value of the acquisition time must be chosen according to the number of cycles required to obtain a good signal-to-noise ratio but an upper limit is imposed by the dynamics of the process that will be characterized by this technique and the drift of the resonance frequency due to temperature changes (9 Hz/ $^\circ$C in this case). The Fourier integral of the signal is $\dfrac{\sin[\pi(f - f_m)T]}{\pi(f - f_m)}$ with its maximum at f_m. If f_m is not an integer multiple of the frequency spacing determined by the FFT, which is $1/T$, the profile of the resonance peak is not precisely measured. In order to overcome this difficulty the HRFFT method discussed above was applied.

The amplitude of the frequency spectrum at the laser beam chopping frequency f_m is stored and the procedure is repeated for different repetition rates around the cell's resonance with steps adequate for a good definition of the maximum, depending on the cavity Q-factor. The recorded amplitude and the frequency values determine the PA peak, which shows a Lorentzian profile, as expected.

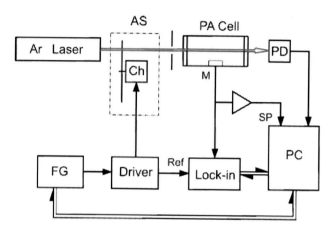

Figure 14. Experimental setup for simultaneous measurement of a PA signal with the lock-in and the PC-based system (PD: power meter, Ch: chopper, FG: function generator, SP: sound port, AS: acoustic shield, M: microphone).

For comparison, the microphone signal for each frequency was processed by a DSP lock-in amplifier (Stanford Research Mod. 830) synchronized with the chopper and in parallel fed into the PC through the sound port, after amplification by a Tektronix AM502 module (Figure 14). The PC system was previously calibrated with a sinusoidal signal.

The whole resonance peak, simultaneously registered at a temperature of 28.6 °C with both systems with a frequency resolution of 1 Hz, is shown in Figure 15 for 630 ppmV NO_2 in pure air. The temperature did not drift more than 0.1°C during the measurement. In order to compare the performance of the systems for application in pollution measurements, they were used with similar integration times. The time constant of the lock-in system was set at 1 second, as well as the acquisition window T in the PC sound card based system. The results show a perfect coincidence between both acquisitions.

The ultimate sensitivity was obtained by determining the mean value of the signal amplitude (S_{air}) around the resonance f_0 plus twice the standard deviation σ, with the laser beam passing through the cell containing pure air at atmospheric pressure. These determinations correspond to 50 ppbV for the sound card-based system and 130 ppbV for synchronic detection. With a random noise, we made certain that this difference is due to the different bandwidths of the instruments for our acquisition conditions.

The system for PA detection based on signal acquisition by a PC sound card is an interesting alternative to synchronic detection. It is of very low cost and allows to develop a compact and portable equipment for field applications, as the whole acquisition and processing system can be included in a laptop computer.

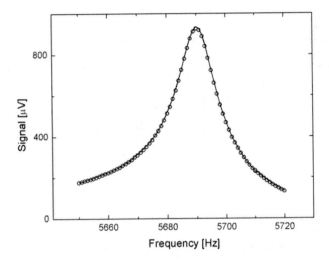

Figure 15. Comparison of the resonance profile acquired by the lock-in (dots) with the scan by the sound card-based system (solid line).

In the specific case of NO_2 traces PA detection, based on solid-state lasers, this system may be included with no harm to the performance already obtained with a lock-in amplifier.

4.2. Excitation by a Mechanically Modulated Green Laser

In order to afford a totally compact and cheap PA system for field applications, we tested the performance of a setup based on an economic and small-sized green CW laser [29].

A sketch of the setup used along the measurements with this laser is shown in Figure16. The acoustic cavity is the same as in section 3.3.1. The laser is a compact diode–pumped Nd:YAG device emitting more than 50 mW at 532 nm (beam diameter $\approx$ 1.5mm). The output polarization direction was defined by a polarizing cube and the laser power was monitored through the measurements. Thus, the average power incident on the cell was around 15 mW with a stability of circa 10%. For very low concentrations, double passage of the laser radiation through the cell was used. In order to get a precise scanning of the resonance profile, the mechanical chopper was stabilized within 0.2% at 1 kHz through feedback from the computer connected to the lock-in amplifier. In each measurement we performed a scanning over the whole resonance to determine the maximum amplitude, which drifts because of temperature changes. The signal for each frequency was acquired with a lock-in time constant set to 3 sec and twice averaged. Simultaneously, the power of the incident laser radiation was recorded and each measurement was the result of the normalization of the signal to the mean value of the laser power.

Errors in the determination of the resonance amplitude may have origin in fluctuations of the modulation frequency, acoustical or electrical noise and fast variations of the laser power. We studied the precision of the measurements of the amplitude to mean laser power ratio by repeating the scanning of the resonance several times for a fixed sample pressure. The resulting standard deviation was lower than 5% of the maximum amplitude for mixtures of low and high NO_2 concentrations.

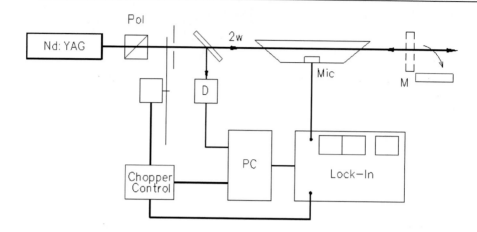

Figure 16. Experimental PA setup with a CW green laser as the excitation source.

An important source of error may also stem from the manufacturing of gas mixtures, mainly at low concentrations. We repeated twice the procedure for preparation of the 5 and 140 ppmV mixtures and verified that in both cases the reproducibility of the signal amplitude was within 10 %. Thus, we inferred that the main source of error in determining the calibration curve consisted of the uncertainties in the concentration values.

The signal from 45 ppmV NO_2 in pure air at atmospheric pressure and the fitting by a Lorentzian profile are shown in Figure17; at the same time the background with and without laser radiation propagating through a pure air sample, at atmospheric pressure also, are shown. The background signal is practically due to intrinsic acoustical and electrical noise of the microphone.

The calibration of this system is performed with mixtures between 2 and 1500 ppmV of NO_2 in chromatographic air at atmospheric pressure. The value of the PA signal amplitude obtained for pure air with laser irradiation lets us conclude that, on the basis of SNR=1, the minimum concentration that could be measured with this method is about 20 ppbV.

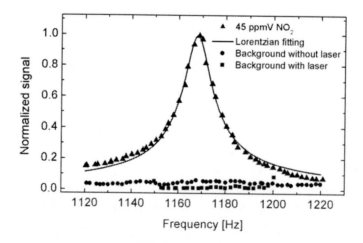

Figure 17. PA signal from a NO_2/air sample and background from pure air.

4.3. Excitation by a High Repetition Pulsed Green Laser [30]

The maximum amplitude of the resonant PA signal from NO_2 in high Q cells is not easily determined with mechanically amplitude-modulated laser systems due to the jitter of the chopper. We propose the use of an acousto-optically Q-switched laser with low phase jitter instead.

The experimental setup is based on a home-made [31], diode pumped, acousto-optically Q-switched Nd:YAG laser with intracavity second harmonic generation (λ=532 nm, power up to 40 mW at 5 kHz, power stability better than 10%, pulse-width 40 nsec FWHM, beam diameter ~ 2mm). The laser, whose jitter is about 0.01% at 3 kHz, is operated at a repetition rate that allows resonant excitation of the first radial mode of the cylindrical closed Pyrex resonator already used in section 3.3.2. The acoustic signal, obtained from the scanning through the resonance peak at ~5700 Hz in steps of either 0.5 or 1Hz, is synchronously detected by the microphone and the lock-in amplifier set at time constants of 1 or 3 sec, depending on the S/N ratio, and twice averaged in a computer. The computer-controlled lock-in amplifier drives the acousto-optical modulator. In each measurement scanning over the whole resonance, which drifts in frequency because of temperature changes or mixture composition, allows to determine the maximum amplitude. The results are stored in the computer simultaneously with the laser power meter (Laser Probe Rk3100) readings for normalization purposes (Figure 18).

The system was calibrated with mixtures prepared at our lab by successive dilutions starting from a concentration of 1300 ppmV down to 7 ppmV. A good correlation (0.99) was obtained for the linear regression applied to the amplitude at constant input power vs concentration. From the repetition of the signal amplitude measurement for 30 ppmV, over one hour and through new mixtures, an error of the PA signal of 5 % was deduced.

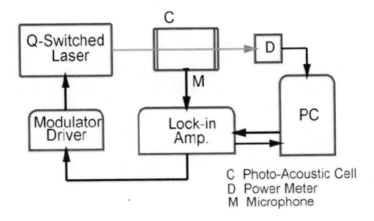

Figure 18. Experimental setup for PA measurements with a high repetition Q-switched laser.

A detection limit of around 50 ppbV was determined for samples of pure air at atmospheric pressure when laser radiation crosses the cell. This value is the result of the signal amplitude average over the frequency range around the resonance plus twice the standard deviation.

The influence of water vapor in NO_2 samples on the PA signal was studied. The comparison between the determinations from samples of NO_2 in pure air at atmospheric pressure with mixtures in outdoor air clearly shows a significant lowering of concentration over time (60% in one hour) in the latter case. By filtering the humid outdoor air through 150 mg of anhydrous $CaSO_4$ the PA signal from this mixture shows the same change over time as NO_2-pure air.

The gases from car exhausts are mainly CO, CO_2, NO_x, hydrocarbons and water. Among all these molecules, NO_2 is the only one that presents intense absorption bands in the visible range. This fact encouraged us to try the application of this system to the measurement of NO_2 content in car exhausts.

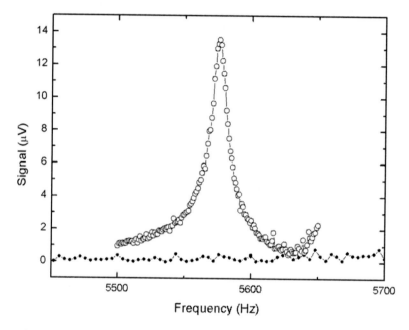

Figure 19. PA signal from a sample from a car exhaust. Open circles: cold catalytic converter. Dots: hot catalytic converter.

Samples of different cars at idling operation, drawn into our PA cell directly from inside the exhaust pipe through two filters for 0.22 μm particles and, in some cases, through the desiccant, were transported to our laboratory for immediate analysis. The filters were added because of the high content of particles in diesel exhausts. In our experiment we compared the content of NO_2 from different car exhausts without time delay after the sample extraction in order to guarantee neither NO_2 removal because of the presence of water nor formation from NO. The exhaust gases from gasoline and diesel cars with catalytic converters were analyzed at both normal temperature of engine operation and just after ignition, before the conversion process takes place. In Figure 19 the results from a gasoline-powered car, with a displacement of 1600 cm^3, are shown when the catalytic muffler is hot and cold. The concentration of NO_2 reduces from 4 ppmV a couple of minutes after the beginning of operation of the engine to a level under the minimum measurable concentration, once steady temperature is reached.

The results of NO_2 emissions obtained for different types of cars are summarized in Table 1. The values reflect the NO_2 content immediately after collecting the sample.

The NO_2 concentration, with and without desiccant for the same car, is similar, as expected according to the specifications about retention of other gases for anhydrous calcium sulfate from catalogue. In all cases, the NO_2 content of diesel exhausts is higher than that of gasoline cars. The catalytic converters of gasoline automobiles reduce drastically the amount of NO_2 in samples taken from their exhaust pipes. On the contrary, new diesel-powered cars with catalytic converters different than those applied to gasoline engines, show no NO_2 elimination. Compared to these determinations on diesel cars, the methane gas engines, despite being old models, show rather low pollutant emission.

Table 1. PA analysis of NO_2 emissions in different car exhausts.

Type of engine								
Gasoline	A Mod.99	B Mod.97	C Mod.91					
Natural gas				D Mod.90	E Mod.93			
Diesel						F Mod.93	G Mod. 2000	H Mod. 2000
NO₂ content [ppmV]								
Without CC	4	1	2	3	0.5	28	45	9
With CC	0	0	---*	---	---	---	36	9

* indicates absence of catalytic converter

5. ONE-DIMENSIONAL PIPE RESONATORS

5.1. Acoustic Transmission Line Model

We have shown above that, in our systems, the PA signal from pure air is the same as the signal that is obtained when the laser is blocked. This is characteristic of experiments with excitation by visible radiation in cells ended in optical glass windows. When infrared radiation is used, instead, windows heating is important. It is synchronic with the excitation and sometimes imposes too high detection limits. In this case special geometries of acoustical cavities are used, with filters that may be in form of buffers at the ends of the resonator. The dimensions of this type of acoustical cavities are such that the main resonances correspond to longitudinal modes.

The frequency response of a PA cell of such characteristics has been simulated [32]. In particular, a model is applied to the design of a flow-through acoustic resonator with buffers. Also, gas inlet and outlet openings are placed at the nodes of the second longitudinal mode (the first non-zero mode at the microphone in the center of the cell), for avoiding noise due to turbulence in the flow regime. In Figure 20 the transversal view and the dimensions of a high-density polyethylene flow-cell which was manufactured at our lab, are shown. The design exhibits the characteristics named above.

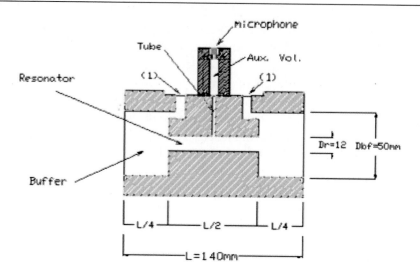

Figure 20. Transversal view of the plastic cell with a Helmholtz volume containing the microphone (1): gas inlet-outlet.

An acoustic resonator may be considered as a one-dimensional one when the sound wavelength is greater than the transversal dimensions [27,32]. An adequate description is based on the analogy between the one-dimensional acoustic equations and the electromagnetic equations of a transmission line. The lossless equations of an acoustic system may be written as:

$$\frac{\partial p}{\partial t} + \frac{\rho c^2}{S} \frac{\partial u}{\partial x} = 0 \qquad (18)$$

$$\frac{\rho}{S} \frac{\partial u}{\partial t} + \frac{\partial p}{\partial x} = 0 \qquad (19)$$

where p is the pressure, u the flow velocity, ρ the gas density, S the transverse section of the resonator and c the sound velocity. From the analogy with the equations for transmission lines, one can deduce the expressions for the acoustic capacitance and inductance of a duct of unit length: $C^* = S / \rho c^2$ and $L^* = \rho / S$. Losses, as we have seen before (see section 3.3.1), happen mainly at the surface; they are introduced as a resistance of the cable and conductance of the dielectric. The relationship between the pressure of the sound wave at the input and output of a duct of uniform cross section and length dx may be represented by an equivalent T circuit, such as the unitary network included in Figure 21. The transition between ducts of different diameters, as in a resonator ended with buffers, is represented by series impedance determined by the dimensions of both sections. The current source dI represents the heating produced by absorption of laser light and, in the case of traces detection, a uniform distribution of these sources along the whole resonator is a good approximation. In the model, a correction is introduced in both open endings of the resonator as a small lengthening, which may be interpreted as a contour effect due to the passage from a one-dimensional duct (resonator) to the three-dimensional buffer volumes. The microphone is enclosed in an additional volume (Helmholtz cavity) connected to the main resonator through a straight narrow tube (3mm dia.). In Figure 21 the equivalent circuit is shown, which represents the unit network of the resonator and the transmission line corresponding to the tube. The Helmholtz cavity is represented as termination impedance in parallel with the microphone's one.

The experimental frequency-response of the cell is studied under a resonant regime using the high repetition green laser named in section 4.3, glass windows (no window heating) and NO_2-N_2 mixtures. The results are compared with the modeled signal, taking into account the typical frequency response of the microphone (Knowles EK3024), as given by the manufacturers. The sole parameters of the model are the complex impedances representing the microphone; the starting values upon running the program are taken from Bijnen et al.[34] and the model appeared to be mildly sensitive to the microphone's capacitance value.

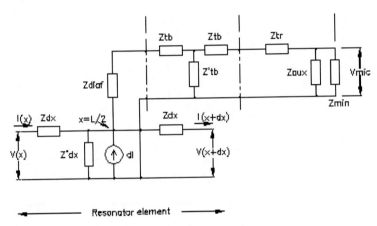

Figure 21. Electric equivalent circuit of the resonator of the plastic PA cell with the Helmholtz volume connected through a transmission line.

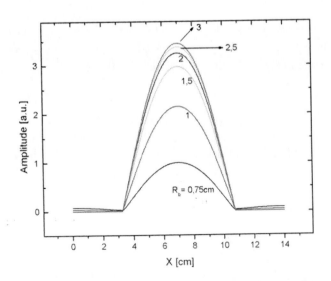

Figure 22. Calculated amplitude of the pressure wave along the cell axis for different buffer-to-resonator diameter ratios.

The calculated PA signal at the center of the cell reaches its optimum value for a ratio buffer-to-resonator diameter around 3 (Figure 22).

Two Helmholtz cavities of different lengths were attached to the main resonator for checking the accuracy of the model with this particular geometry. The results are shown in Figure 23, where the peaks correspond to two different Helmholtz volumes. Good agreement between model and experiment was obtained.

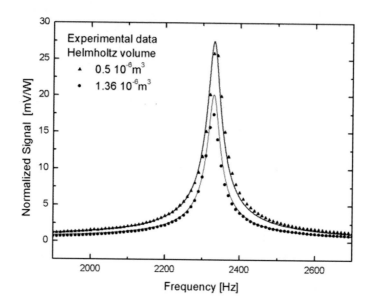

Figure 23. Results from model (line) and experiment for two different Helmholtz volumes.

5.2. SF$_6$ Multiphoton Absorption Cross Section Determination [35]

The precise knowledge of the PA spectroscopy of NO$_2$ as well as its spectral characteristics allow using the PA technique as a calibration method for determining unknown spectral parameters of other molecules. Here we describe how the multiphoton cross section of SF$_6$ is measured that way.

Pulsed photoacoustics using a TEA CO$_2$ laser is applied to the detection of SF$_6$ traces in N$_2$ at atmospheric pressure as part of an experiment that intends to measure low concentrations of tracer gases for studying gas diffusion in plants. A home-made TEA CO$_2$ laser is used with a special cavity design, based on a diffraction-coupled resonator, which allows obtaining a spatially homogeneous and well collimated beam at a large distance from the laser (ca. 7m).

The amplitude of the resonance peak corresponding to the n-th eigenmode of an acoustic cavity at the υ_n frequency for a pulsed system is proportional to the absorbed laser energy in accordance to an expression alike (17):

$$S_n(\upsilon_n) = K_n \ \sigma(\Phi) \ \Phi C \qquad\qquad (20)$$

where K_n (setup constant) is the factor which depends on the overlapping integral of the n-mode pattern and the laser beam intensity distribution, C is the SF_6 concentration and σ the absorption cross section. The latter is indicated as dependent on laser fluence, $\sigma_{eff} \equiv \sigma$ because we deal with a multiphotonic absorption (MPA) process in SF_6; moreover, it is an unknown magnitude for our experimental conditions (samples with 700 Torr N_2).

To determine the values of the MPA cross section of SF_6 at 947.7 cm^{-1} we analyze the results comparing with PA studies on NO_2 traces in N_2, excited by the pulsed visible laser, in the same cell as the SF_6 experiment. A plastic cell with a small radius resonator and buffers at both ends was used, in order to minimize the signal coming from the heating of the infrared transparent windows. It seemed that assuming the same setup constant in both cases was reasonable due to two main reasons: on one hand, the spectral power distribution which was obtained for both infrared and visible systems was identical, and, on the other hand, in both systems a very efficient energy transfer to translation takes place during the laser pulse with rates higher than 5.10^8 s^{-1}[9,36]. So, the setup constant was determined from the slope of the calibration line for the NO_2-N_2 system (energy-normalized spectral amplitude of the (002) resonance vs absorbing species pressure). For samples concentration from 18 to 660 ppmV the line's correlation coefficient was 0.999 and the slope error 1%.

In the SF_6 system, the PA signal from samples of four particular dilutions in N_2 (between 1 and 6 ppmV) was acquired for six different laser energy values each, in the range 5 to 200 mJ/cm^2. The (002) peak's amplitude is related to σ_{eff}, laser energy and SF_6 concentration by the setup constant K_n deduced above for NO_2. The NO_2 absorption cross section for this calculation was $(1.46\pm0.03).10^{-19} cm^2$, measured at our lab with the laser used along the PA measurements (linewidth = 1 cm^{-1}) itself. The error of σ_{eff} is estimated around 20%, where the greatest contribution is due to the pyroelectric detector (~10% by the manufacturer). When the expression $<\eta> = \sigma_{eff}\Phi/h\upsilon$, with $h\upsilon$ the photon energy, is applied, we obtain the curve on Figure 24, at the same time as $\sigma_{eff}(\Phi)$.

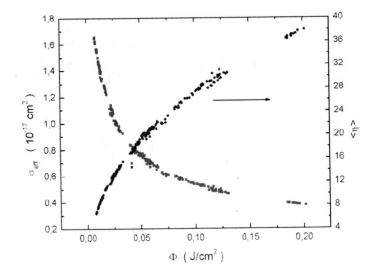

Figure 24. Multiphoton absorption cross section of SF_6 in N_2 and average number of absorbed photons per molecule.

$<\eta>$ shows a dependence on laser energy of the Φ^{β} form, where the coefficient β determined from the non-linear fit of the experimental data is 0.516±0.003.). The magnitude of $<\eta>$ shown in Figure 24 are consistent with values obtained by other authors [37], under different collisional regimes.

6. CONCLUSION

In this chapter different results, from the academic and practical point of view, related to PA detection of NO_2 traces with visible lasers are resumed. All the experiments described therein have been performed at the Spectroscopic Techniques division of the CEILAP (CITEFA-CONICET) at Buenos Aires.

Diverse features, even difficulties, are discussed. In particular, the Q-factor shows dependence on the total pressure, so the samples from urban air must be taken under the same conditions as those applied during the calibration process. Moreover, filtering by a desiccant is necessary when non-flowing samples are measured, so that water vapor might be eliminated to avoid its influence on the measurements. One problem that appears is the phenomenon of adsorption of NO_2 molecules at the walls of the PA reservoir. It is recommendable to perform low concentrations measurements under flow regime in order to obtain greater precision. On the other hand, the PA technique itself is useful for studying the adsorption process.

The processing method called *HRFFT* represents a powerful tool for a precise determination of the amplitude of the PA signal and it is used in several experiments.

The PA systems described above are used for testing car exhausts gases and controlling catalytic converters. The setup may be simplified by building a low-cost PA arrangement of signal detection and processing based on an audio amplifier and a laptop. This way, neither lock-in amplifiers nor expensive A-D cards or digital oscilloscopes are necessary.

With the systems shown above, the lowest measurable concentration of NO_2 in air is about some tens of ppbV. This limit allows detecting pollution in urban areas.

The one-dimensional model predicts the behavior of acoustic cavities suitable for flow regime even with special designs, as Helmholtz cavities, where the microphone is far from the resonator. This is useful for optimization of cell designs.

Resuming, the application of solid-state lasers and simple acquisition systems to photoacoustics, as shown in this chapter, allows to conceive compact and portable systems for field or *in-situ* measurements. Furthermore, the development of lower cost laser diodes emitting in the blue region or even powerful LEDs is growing very fast. It represents an interesting alternative to our devices and it is being currently studied at the CEILAP.

ACKNOWLEDGMENTS

This work was performed with finance aid from CONICET and the ANPCyT. I wish to thank all my colleagues who contributed in a very important way to this project with technical assistance and useful discussions.

REFERENCES

[1]   Durry, G.; Megie, G. *Appl.Opt*. 1999, 38, 7342-7354.

[2]   Sigrist, M.W. in Air monitoring by spectroscopic techniques; Sigrist, M.W.; Ed., *Chemical Analysis* **127**; John Wiley & Sons, New York, 1994; pp. 163-238.

[3]   Sigrist, M.W. *Rev.Sci.Instrum*. 2003, 74, 486-490.

[4]   Webber, M.E.; Pushkarsky, M.B.; Patel, C.K.N. SPIE's International Symposium on Optical Science and Technology, 7-11 July 2002, Seattle, WA, *Diode Lasers and Applications, Paper Number* **4817**-11.

[5]   Rosencwaig A., Photoacoustics and Photoacoustic Spectroscopy, *Chemical Analysis* **57**, John Wiley & Sons, New York, 1980.

[6]   Schäfer, S.; Miklós, A.; Hess, P. in *Progress in Photothermal and Photoacoustic Science and Technology*; Mandelis, A. ;Hess, P. Eds.; SPIE, Bellingham, Wash., 1997; Vol. 3, pp. 254-288

[7]   Merienne M.; Jenouvrier, A.; Coquart B. *J. Atmos. Chem*. 1995, 20, 281-297.

[8]   Gillispie, G.D.; Khan A.U. *J. Chem. Phys*. 1976, 65, 1624-1633.

[9]   Okabe, H., *Photochemistry of Small Molecules*; John Wiley & Sons, New York, 1978; pp. 227-235.

[10] Haaks, D.; Schurath U. in Laser-induced Processes in Molecules; Kompa, K.L.; Smith, S.D., Eds.; Springer-Verlag, Berlin, 1979; pp.352-355.

[11] Vandaele, A.C.;Hermans, C.; Simon, P.C.; Carleer, M.; Colin, R.; Fally, S.; Merienne, M.F.; Jenouvrier, A.; Coquart, B. *J. Quant. Spectrosc. Radiat. Transfer* 1998, 59,171-184.

[12] Schneider, W.; Moortgat, G.K.; Tyndall, G.S.; Burrows J.P. *J. Photochem. Photobiol*. A, 1987, 40, 195-217.

[13] Bernegger, S.; Sigrist, M.W. *Infrared Phys*. 1990, 30, 375-429.

[14] Angus, A.M.; Marinero, E.E.; Colles M.*J. Opt. Commun*. 1975, 14, 223-225.

[15] Claspy, P.C.; Ha, C.; Pao, Y. *Appl. Opt*. 1977, 16, 2972-2973.

[16] Koch, T.; Fenter, F.; Rossi, M. *Chem. Phys. Lett*. 1997, 275, 253-260.

[17] Brand, C.; Winkler, A.; Hess, P.; Miklós, A.; Bozóki, Z.; Sneider, *J. Appl. Opt*. 1995, 34, 3257-3266.

[18] Karbach, A.; Hess P. *J. Chem. Phys*. 1986, 84, 2945-2952.

[19] Karbach, A.; Hess, P. *J. Chem. Phys*. 1985, 83, 1075-1084.

[20] Miklós, A.; Lörincz A. *Appl. Phys. B* 1989, 48, 213-218.

[21] Slezak, V. *Rev. Sci. Instrum* 2003, 74, 642-644.

[22] Brigham, E.O.; The Fast Fourier Transform and its Applications, Prentice-Hall, New Jersey,1988.

[23] Slezak, V. *Appl. Phys. B* 2001, 73, 751-755.

[24] Ferreira, M.; Herscovich Ramoneda, E.; Slezak, V.; Peuriot, A.; Santiago, G. *Anales Asociación Física Argentina* 2004, 16, 62-65.

[25] Schmohl, A.; Miklos, A.; Hess, P. *Appl. Opt*. 2001, 40, 2571-2578.

[26] Henningsen, J.; Melander, N. *Appl. Opt*. 1997, 36; 7037-7045.

[27] Miklós, A.; Hess, P.; Bozóki, Z. *Rev. Sci. Instrum*. 2001, 72, 1937-1955.

[28] Santiago, G.; Slezak V.; Peuriot, A. *Appl. Phys. B* 2003, 77, 463-465.

[29] Slezak, V.; Santiago, G.; Peuriot, A. *Optics and Lasers in Engineering* 2002, 40, 33-41.

[30] Slezak, V.; Codnia, J.; Peuriot, A.L.; Santiago, G. *Rev.Sci.Instrum.* 2003, 74, 516-518.

[31] Codnia, J.; Azcárate M. L. *Anales Asociación Física Argentina* 1999, 11, 80-83.

[32] Bernegger, S.; Sigrist, M.W. *Infrared Phys.*1990, 30, 375-429.

[33] Slezak, V.; Peuriot, A.; Santiago, G.; Codnia *J. Proceeding 5th Iberoamerican Meeting on Optics and 8th Latin American Meeting on Optics*, Lasers and their Applications (RIAO 2004 OPTILAS), SPIE electronic edition, 5622, CDS154.

[34] Bijnen, F.G.C.; Reuss, J.; Harren, F.J.M. *Rev.Sci.Instrum.* 1996, 67, 2914-2923.

[35] Slezak, V.; Peuriot, A.L.; Santiago G. *Journal de Physique IV* 2005, 125, 39-41.

[36] Kolomilskil, Yu.; Marchuk, V.S.; Ryabov E.A. *Sov.J.Quantum Electron.* 1982, 12, 1139-1142.

[37] Lyman, J.L.; Quigley G.P.; Judd O.P. in *Multiple-photon excitation and dissociation of polyatomic molecules*; Cantrell C.D.; Ed.; Springer-Verlag, Berlin, 1986; pp.9-94.

In: Perspectives in Optics Research
Editor: Jeffrey M. Ringer, pp. 83-170

ISBN: 978-1-61122-934-9
© 2011 Nova Science Publishers, Inc.

Chapter 5

CHEMICAL OXYGEN IODINE LASER: CURRENT DEVELOPMENT STATUS AND APPLICATIONS

*R. Rajesh, Gaurav Singhal, P. M. V. Subbarao, Mainnudin, R. K. Tyagi and A. L. Dawar**

Laser Science and Technology Centre,
Metcalfe House, Delhi-54, India
Indian Institute of Technology, Delhi, 16, India

ABSTRACT

Chemical Oxygen Iodine Laser (COIL) has emerged as one of the most promising and powerful lasers in the recent past because of its immense potential in wide ranging applications in material processing and defense. The low temperature, low pressure COIL active medium provides low divergence and excellent beam quality. The wavelength of the laser (1.315 micron) is compatible for transmission through fiber and atmosphere. COIL system can be scaled up to several megawatts and can also be operated in a wide range of the pulse repetition regimes from Hz to GHz range. These features clearly indicate their potential in the industry for cutting and welding of materials, remote applications like cutting and dismantling of obsolete nuclear reactors or underwater cutting and in military applications as antimissile weapon.

Since its invention in 1978, the COIL research has passed through many stages and the system development has now reached to such a maturity level where advanced countries are planning deployable systems based on this laser. The COIL systems of hundreds of kilowatt power have been realized. Further efficiencies of the order of 35 % are becoming possible with the realization of supersonic flows in the cavity.

The article briefly presents the history of COIL along with the detailed review on design and development status of various critical subsystems like singlet oxygen generator, supersonic nozzle geometries, cavity configurations, and schemes for atmospheric discharge COIL. Various applications of COIL in civil and military

---
* E-mail address: a_l_dawar@yahoo.com (Corresponding author)

including Airborne Laser, Tactical High Energy Laser, material processing, remote dismantling of nuclear reactors and rock crushing for oil exploration are also discussed.

1.0. BACKGROUND

The development of lasers in general, and high power lasers in particular, has been a turning point in the history of science and engineering. It has produced a completely new type of systems with potentials for applications in a wide variety of fields. During sixties, lot of work had been carried out on the basic development of almost all the major high power lasers including gas dynamic and chemical lasers. Almost all the practical applications of these lasers in defense as well as in industry were also identified during this period. The motivation of using the high power lasers in strategic scenario was a great driving force for the rapid development of these high power lasers. In early seventies, megawatt class carbon dioxide gas dynamic laser was successfully developed and tested against typical military targets. The development of chemical lasers took slightly longer time because of involvement of multidisciplinary approach. Though Kasper and Pimental [1] reported the first chemical laser based on flash photolysis of iodine, the major research and technical development, however, could take place in seventies only. During a period of ten years from 1970 – 80, chemical lasers particularly hydrogen fluoride, deuterium fluoride matured to an extent that megawatt class system such as Middle Infrared Advanced Chemical Laser (MIRACL) were developed for various military applications.

The major emphasis has been on the development of shorter wavelength CW or high average power pulsed lasers mainly for defense applications. However, with the development of these lasers, the area of material processing has immensely benefited. The most established applications are the laser welding and laser machining including cutting, drilling and shaping. Both for defense as well as material processing applications, it is advantageous to use shorter wavelength and excellent beam quality in order to have high energy density at the desired place.

Chemically excited iodine lasers are the second-generation lasers in this category because of their superior salient features over the HF / DF. COIL is the first of its kind with electronic transition, which was first demonstrated by McDermott [2] in 1978 at Air Force Research Laboratory (AFRL), USA. The lasing is achieved between the electronically excited level of iodine atoms $I(^2P_{1/2})$ and its ground level $I(^2P_{3/2})$ with stimulated emission at 1.315μm . It is worth mentioning that the lasing from iodine atoms was initially demonstrated, way back in 1964, by Kasper and Pimentel [1] employing UV photolysis of certain alkyl iodide compounds. Later on in 1966, Demaria and Ultee [3] were able to scale up this photolytic iodine laser with flash lamps and produce pulse energy of about 65 Joules. Even though these photolytic iodine lasers have the potential of yielding high-energy laser pulses, the CW operation is limited due to the non-availability of the UV sources for continuous pumping. The quest for high power in CW mode had motivated the research in alternative pumping techniques for these molecules that resulted in the invention of the Chemical Oxygen Iodine Lasers (COIL). It uses chemically generated, electronically excited oxygen molecules (singlet oxygen), $O_2 (^1\Delta_g)$, as the pumping source.

Some of the important features of COIL that has prompted the rapid development of this laser can be summarized as:

- Higher energy (J/gm) as compared to HF and DF
- Comparatively non-toxic.
- Low temperature – low-pressure laser.
- Can be very compact and light
- Extremely high chemical efficiency ~ 0.3 [4,5,6]
- Excellent beam quality
- The laser output is proportional to the flow of the laser and pumping medium and hence the power scale up is linear.
- Since the laser is under electronic transition, the oscillating wavelength is at near IR region ($\lambda = 1.315$ μm), which is fiber compatible.
- Relatively better laser material interaction due to its shorter wavelength.
- More than 200 kW continuous wave power modules have already been demonstrated by AFRL [7]. This potential has led to the use of COIL in various military and civilian applications.
- The possibility of carrying high power beams via optical fibers makes it extremely useful for decommissioning and dismantling of dangerous structures like obsolete nuclear reactors through remote control. [8]

From the first realization of this laser, there has been significant progress in the development of this laser because of worldwide interest in the shortest wavelength and high efficiency. With the introduction of supersonically flowing (Supersonic COIL) laser gas (i.e., singlet oxygen and iodine gas mixture) there is a large potential for power scaling possibility. The gain in the cavity is inversely proportional to temperature and thus the adiabatic expansion of the lasing medium helps in higher gain. In earlier COILs where the medium used to flow at subsonic speed not only the efficiency used to be significantly lower [9,10,11,12] but also the performance was also very much sensitive to the iodine concentration [13,14] Advanced COILs, on the other hand, are supersonic COILs [8,15,16, 17] (SCOIL), which are compact and have much higher efficiencies. Further the output power is fairly constant in a wide range of iodine concentration in these systems.

Even to day the most advanced COILs are based on the same principles as suggested by McDermott, except that the techniques for producing singlet oxygen have improved from bubbler generator to jet generator. Efforts are also being made to make the system more efficient as well as alternatives are being tried to dispense with Basic Hydrogen Peroxide and chlorine for the production of singlet oxygen. These include the concept of electri- COIL [18] and All Gas Iodine Laser (AGIL) system approach [19].

The first demonstrated COIL power at US Air Force laboratory was only few mill watts and over a short span of 27 years, modules of 200 kW power have been developed to push the power levels to a megawatt class. COIL technology has undergone numerous improvements and chemical efficiencies as high as 33% have been already demonstrated [5]. In addition, COIL power was also demonstrated under various modes like Q-switching, Mode locking, Frequency doubling etc. The major steps of advances or breakthroughs in COIL research are given in Table (1).

Table 1. Stages of COIL research

Year	Organization/ Country	Power (W)	η (%)	Remarks
1978	AFRL, USA	0.004	1×10^{-5}	First COIL, Horizontal flow, Subsonic
1981	TRW, USA	2000	11	First kW level, Subsonic
1982	AFRL, USA	4600	8	Subsonic
1982	VNIIEF, Russia	0.01		First reported work from Russia, Subsonic
1982	Ben Gurion U, Israel	5		First reported work from Israel,
1983	Natl. Defense Acad, Japan	10		First reported work from Japan,
1984	ONERA, France	4		First reported work from France, Subsonic
1984	AFRL, USA	1600	12	First Supersonic COIL
1984	TRW, USA	4200	15	Supersonic
1988	DICP, China	0.130		First from China, Electrical discharge
1989	AFRL, USA	39000	24	Supersonic, ROTOCOIL, Max. Reported power with rotating disc SOG
1989	AFRL, USA			Frequency doubled
1989	Lebedev, Moscow, Russia	5		Photo-dissociation of Ozone
1990	AFPL, USA	630		Magnetically Q-switched COIL
1991	Lebedev, Samara, Russia			Invention of jet generator
1991	Institute of Physics, Czech Rep.	58	5	First reported work from Czech Rep. Subsonic
1995	RD, USA	17500	21	Supersonic, mode limited aperture
1996	AFPL, USA	1750	28	Compact supersonic, VERTI-COIL
1996	AFPL			Mode locked COIL
1997	Lebedev, Samara, Russia	200	22	First Jet generator based COIL, Horizontal geometry
1997	DLR Germany	5000		First reported work from Germany
2000	Lebedev Samara, Russia	1400		Jet generator based Vertical geometry
2000	Samara, Russia	700 W		Development of first ejector nozzle applicable to COIL
2000	AFRL, USA	~1W		Demonstration of first All gas iodine laser, with $NCl(^1\Delta)$ replacing $O_2(^1\Delta)$
2001	Ben Gurion U, Israel		30	Development of nozzles employing transonic and supersonic Iodine injection
2002	Boeing, USA	20000	22	Nitrogen dilution, With ejector nozzle with tabs, Cross flow JSOG
2003	LASTEC, India	350	17.5	First reported work from India, Jet generator with Angular geometry, conventional Supersonic nozzle
2003	RILP, Russia	25 mJ, 50 µs,		Lasing of first Fullerene Oxygen Iodine Laser (FOIL)
2003	AFRL, USA	15		Multi-watt All Gas Iodine Laser (AGIL)
2003	Ben Gurion U, Israel		33	Development of nozzles employing transonic and supersonic Iodine injection
2004	Tokai University	599W	32.9	X-wing Nozzle
2004	TRW/ Boeing	200000		The highest power COIL existing in world
2005	CU Aerospace, USA	0.220		Lasing of Electric COIL with rf produced $O_2(^1\Delta)$

2.0. Basics of Chemical-Oxygen Iodine Laser

The chemical Oxygen Iodine Laser operates on the hyperfine components of the $5^2P_{1/2}$ - $5^2P_{3/2}$ transition in atomic iodine while chemically pumped from metastable $O_2(^1\Delta_g)$. The principle of this laser can be described by the following set of chemical reactions involved in the lasing process [2]:

$$O_2(^1\Delta) + I(^2P_{3/2}) \Leftrightarrow O_2(^3\Sigma) + I^*(^2P_{1/2}) \qquad !\ \text{Pumping} \quad (1)$$

$$I(^2P_{1/2}) + nh\nu \rightarrow I(^2P_{3/2}) + (n+1)h\nu \qquad !\ \text{Lasing} \quad (2)$$

$$O_2(^1\Delta) + I_2 \rightarrow O_2(^3\Sigma) + I^*(^2P_{1/2}) \ !\ I_2\,\text{Dissociation} \qquad (3)$$

The near resonant energy transfer (energy difference of 279cm^{-1}) between the singlet oxygen and iodine atoms is rapid and a near equilibrium between the upper and lower level is quickly established with a temperature dependent equilibrium rate constant $K_{eq}(T)$ [20],

$$K_{eq}(T) = \frac{[I(^2P_{1/2})][O_2(^3\Sigma)]}{[I(^2P_{3/2})][O_2(^1\Delta)]} = 0.75\exp(402/T) \tag{4}$$

Where the square bracket [ ] indicates the corresponding concentration of the species. In COIL operation, in addition to the pumping, the singlet oxygen also serves in dissociating the iodine molecules in to atoms.

2.1. Gain

The energy level diagram of COIL illustrating the resonant energy transfer from this metastable energy reservoir to the upper level of atomic iodine along with various states involved in the dissociation of iodine molecules are shown in Fig. (1). The hyperfine structure and the spectrum of the $^2P_{1/2}$- $^2P_{3/2}$ shown in Fig. (2) clearly indicates that the maximum gain is achieved for the F'=3 to F''=4 resulting in a wavelength of 1.315 μm [20]. The gain (α) during the population inversion can be written as,

$$\alpha = \sigma(g_u N_u - g_l N_l) \tag{5}$$

i.e for the maximum gain case between F'=3 to F''=4 one can write as

$$\alpha = \sigma(g_u[I^2P_{1/2}F'=3] - g_l[I^2P_{3/2}F''=4]) \tag{6}$$

Where, σ is the transition cross-section, N and g represents the population number density and degeneracy of the upper and lower level indicated by the suffixes u and l respectively. Since the degeneracy of a hyperfine state F is given as g=2F+1: the degeneracy of the important states, F'=3 and F"=4 works out to be 7 and 9 respectively. For the case of population in all hyperfine levels, the population densities can be written as

$$[I(^2P_{1/2}, F'=3)] = \frac{7}{12}[[I(^2P_{1/2})]$$

(7)

$$[I(^2P_{3/2}, F"=4)] = \frac{9}{24}[[I(^2P_{3/2})]$$

(8)

Further the fraction of singlet oxygen availability in the laser cavity with respect to the total oxygen content is defined as the singlet oxygen yield Y_Δ:

$$Y_\Delta = \frac{[O_2(^1\Delta)]}{[O_2(^1\Delta)]+[O_2(^3\Sigma)]}$$

(9)

From the above relations one can easily obtain the gain in the medium as,

$$\alpha = \frac{7}{12}\sigma\left\{K_{eq}\left(\frac{Y_\Delta}{1-Y_\Delta}\right) - \frac{1}{2}\right\}[I(^2P_{3/2})]$$

(10)

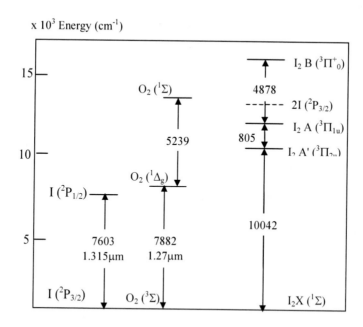

Figure 1. Energy level diagram of O_2, I_2 and I in COIL scheme.

Eqn (10) indicates that the lasing can be achieved even with small quantity of singlet oxygen production. The threshold yield of singlet oxygen at which the lasing can be initiated is obtained by substituting $\alpha=0$ in the above equation,

$$Y_{\Delta,th} = \frac{1}{2K_{eq}+1}$$ (11)

For example, in case of laser medium at room temperature (T=300°K), the threshold yield required is about 15%.

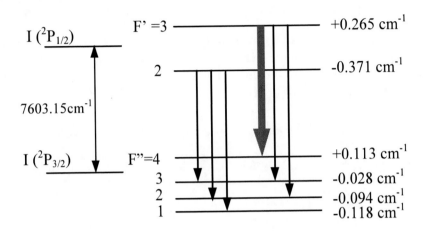

Figure 2. Energy levels and allowed transitions of atomic iodine.

2.2. Power Output and Chemical Efficiency

In COIL, the metastable singlet oxygen molecules are produced in a chemical generator through a two-phase reaction between the chlorine gas and an aqueous Basic Hydrogen Peroxide (BHP) solution governed by the equation [21],

$$Cl_2 + H_2O_2 + 2MOH \rightarrow 2MCl + 2H_2O + O_2(^1\Delta)$$ (12)

The COIL output power is proportional to the amount of pumping medium available for the excitation of iodine atoms at the laser cavity [22],

$$P_{out} = P_{av}\eta_{ext} = 91 \overset{o}{M}_{Cl} U_{Cl}(Y_{\Delta,cavity} - Y_{\Delta,th})\eta_{ext}$$ (13)

Where, P_{av} is maximum available power at the lasing medium in cavity, the factor 91 is the energy in kilo-Joules for one mole of excited iodine atoms. This can also be explained from the energy of the singlet oxygen that one mole of singlet oxygen contains 94.4 kilo-Joules/mole and considering that 4% is spent for the dissociation of iodine molecules to atoms, rest of the energy i.e. 91 kJ is available for excitation of iodine atoms. M_{Cl}, U_{Cl} represent the chlorine gas flow rate supplied into the singlet oxygen generator and the fraction of it utilized in generating the singlet oxygen respectively. $Y_{\Delta,cavity}$ is the singlet oxygen yield at cavity and η_{ext} is the power extraction efficiency.

2.3. Singlet Oxygen Transport Loss

The available yield of singlet oxygen generator depends on various factors apart from its generation efficiency and heuristically one can arrive at,

$$Y_{\Delta,cavity} = Y_{\Delta genexit} - Y_{loss} - Y_{diss}$$

(14)

Here $Y_{\Delta, genexit}$ is the yield at the generator exit, Y_{loss} is the loss of singlet oxygen during its transport from its generator to the laser cavity and Y_{diss} is the singlet oxygen yield utilized for the dissociation of iodine molecules into iodine atoms. The transport loss of singlet oxygen can be estimated for a typical case by considering major loss mechanisms. [23]

$$O_2(^1\Delta) + O_2(^1\Delta) \rightarrow O_2(^1\Sigma) + O_2(^3\Sigma) \quad k_a = 2.7 \times 10^{-17} cm^3/molecule\text{-}sec \quad (15a)$$

$$O_2(^1\Delta) + O_2(^1\Delta) \rightarrow 2\, O_2(^3\Sigma) \qquad\qquad k_b = 1.7 \times 10^{-17}\, cm^3/molecule\text{-}sec \quad (15b)$$

$$O_2(^1\Delta) + wall \rightarrow O_2(^3\Sigma) + wall \qquad (16)$$

$$O_2(^1\Delta) + H_2O \rightarrow O_2(^3\Sigma) + H_2O \qquad k_2 = 4 \times 10^{-18} cm^3/molecule\text{-}sec \qquad (17)$$

Net transport losses can be written as,

$$Q_{loss} = C^o{}_\Delta \left[3 - \left\{ \cfrac{1}{1 + C_\Delta{}^o k_1 \left(\dfrac{x}{V_f}\right)} + \exp\left\{ -\gamma_{ow} v_{th} S\left(\dfrac{x}{V_f}\right) \right\} + \exp\left\{ -k_2 C_{water}\left(\dfrac{x}{V_f}\right) \right\} \right\} \right]$$

(18)

Where,

$C_\Delta{}^o$ -Singlet oxygen concentration at the generator exit

C_{water} -Concentration of water vapor molecules in the flow duct

x   -The transport distance

V_f -Velocity of the flow medium

v_{th} -Thermal velocity of oxygen molecules

S   -The ratio of the transport duct cross sectional area to its volume

γ_{ow} -The probability of heterogeneous relaxation of singlet oxygen with the wall. This accounts for the material properties and machining quality of the transport duct. Here it is considered this value as $\sim 10^{-4}$.

k_1 -The equivalent rate constant of the reactions (15a) and (15b). The value of k_1 has been taken as $2.31 \times 10^{-17}\, cm^3/molecule\text{-}sec$.

The first term in the curly bracket represents the loss factor corresponding to the self-pooling reaction [eqn. (15a and 15b)]. The second one represents the heterogeneous quenching of the singlet oxygen with the wall material of the transport duct [eqn. (16)] and the third one corresponds to the loss due to the water vapor quenching [eqn. (17)].

Therefore for COIL operation, the plenum pressure (transport duct) should be optimized for minimum loss. Figs. (3) & (4) show the calculated values of transport losses as a function of transport distance for typical oxygen partial pressures of 6torr and 8torr in the transport line.

In the above estimation, the water vapor concentration has been taken as 4 % of total oxygen, which is usually the maximum value observed in the experiments. It can be observed from these figures that pooling effects become significantly higher when the oxygen pressure increases from 6torr to 8torr with corresponding transport losses being around 1.25 % and 7 % respectively for a transport length of 10 cm. This also shows that the pooling loss of singlet oxygen is the limiting factor in the high-pressure operation of the COIL. The other important consequence of this pooling loss is the enormous heat released during their quenching and can increase the temperature of the flowing gas medium thus affecting the gain in the laser cavity.

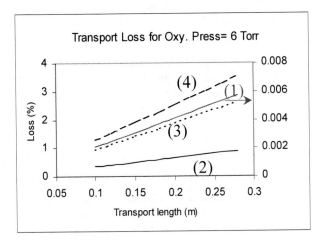

Figure 3. Estimated value of singlet oxygen transport loss with transport distance for the high-pressure case; curve (1): water loss (scale on right axis), curve (2): pooling loss, curve (3): wall loss, curve (4): Total loss.

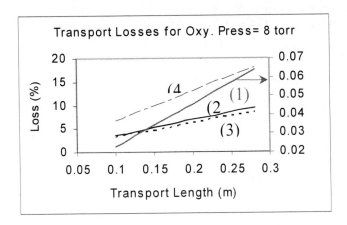

Figure 4. Estimated value of singlet oxygen transport loss with transport distance for the high-pressure case; curve (1): water loss (scale on right axis), curve (2): pooling loss, curve (3): wall loss, curve (4): Total loss.

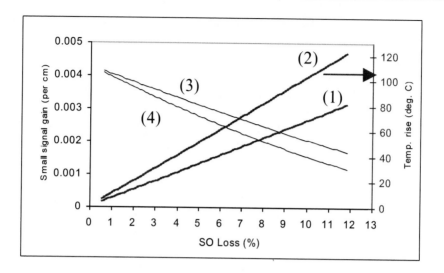

Figure (5). Typical small signal gain variation with SO yield loss for different buffer dilution ration (i.e Cl_2: N_2) of 1:1 and 1:2 (curves 3 & 4 respectively) along with corresponding temperature rise (right axis scale) of the plenum gas (curves 1&2 respectively).

Therefore, the addition of buffer dilution is desirable ($He/Ar/N_2$), which can share the released heat and maintain the temperature rise within an acceptable level and makes efficient transport of the singlet oxygen. Fig. (5) shows the temperature rise of the plenum gas medium and corresponding reduction in the gain obtained at the laser cavity for a typical supersonic operation of COIL with different buffer dilution. It clearly explains the need for the buffer dilution in transporting the singlet oxygen from the SOG to the laser cavity.

In chemical singlet oxygen generators, the reaction occurs between two different phases of reagents and hence water vapor is an unavoidable product. The transport loss estimation shows that the presence of water vapor does not contribute much in the direct deactivation of the singlet oxygen molecules due to their weak reaction rate. However, the deactivation process becomes faster after the mixing of singlet oxygen flow with iodine as it is governed by the fast energy transfer reaction [24]

$$I^*(^2P_{1/2}) + H_2O \rightarrow I(^2P_{1/2}) + H_2O \quad k_3 = 2.0 \times 10^{-12} cm^3/\text{molecule-sec} \quad (19)$$

In that case, the number of singlet oxygen lost per unit volume per second will be,

$$Q_{loss} = (1/3) k_3 [H_2O][I^*] : k_3 = 2 \times 10^{-12} cm^{-3} sec^{-1} \quad (20)$$

It shows that a 2% of water vapor (relative to the singlet oxygen concentration) can make a loss of about 2% in singlet oxygen yield at cavity. Even though in terms of yield loss this looks insignificant, however, the presence of water vapor indirectly affects the performance by quenching the excited iodine as well as by increasing the temperature of plenum in case the iodine concentration is not optimum.

2.4. Iodine Molecular Dissociation

As mentioned earlier, in COIL operation the singlet oxygen is also consumed in dissociation of iodine molecules into iodine atoms that is very much essential for the lasing. The possible dissociation iodine mechanisms believed to occur in COIL system are:

$$I_2 + O_2\,(^1\Delta) \rightarrow I_2^* + O_2\,(^3\Sigma) \quad k_3=7.0\text{x}10^{-15}\ cm^3/molecule\text{-}sec \qquad (21)$$

$$I_2^* + O_2\,(^1\Delta) \rightarrow 2I + O_2\,(^3\Sigma) \quad k_3=3.5\text{x}10^{-11}\ cm^3/molecule\text{-}sec \qquad (22)$$

Therefore, the singlet oxygen yield available at the laser cavity for the population inversion will also be dependent on this fraction utilized for the iodine molecular dissociation (Y_{diss}). For an ideal mixing it can be estimated as [22],

$$Y_{diss} = \frac{N \overset{o}{I_2} F}{\overset{o}{M_{Cl2}} U_{Cl}} \qquad (23)$$

Where N is the number of singlet oxygen molecules required for dissociating an iodine molecule and F is the fraction of iodine molecules dissociated into atoms before they cross the laser cavity. From the energy considerations, N should at least be 2 but practically it is estimated to be about 5 because of the fast deactivation of excited iodine molecules with water vapor present in the cavity:

$$I_2^* + H_2O \rightarrow I_2 + H_2O \quad k_3= 3.0\text{x}10^{-10}\ cm^3/molecule\text{-}sec \qquad (24)$$

It implies that for a typical operation with iodine to chlorine flow rate as 2%, F=0.95 and U_{cl} =0.9, the value of Y_{diss} = 10%. Therefore, the iodine flow rate is a critical factor in determining the COIL efficiencies. In subsonic COILs, the output power is very sensitive to the iodine concentration as compared to that of a supersonic COIL. This is because in subsonic COILs, the residence time of the gas medium in the laser cavity is relatively large and hence if the iodine concentration is higher than a certain critical value, most of the singlet oxygen will be used for the dissociation. Further, the excess of iodine results in reduction of power because of increased deactivation rate of I^* by the non-dissociated molecular iodine;

$$I^* + I_2 \rightarrow I + I_2^* \quad k_3= 3.8\text{x}10^{-11}\ cm^3/molecule\text{-}sec \qquad (25)$$

Various workers have extensively reported the power dependence on iodine dissociation [25, 26]. These studies conclude that the iodine dissociation is dependent on the iodine injection mechanisms and flow parameters, which in turn determine the small signal, gain inside the cavity.

2.5. Extraction Efficiency

In gas flow lasers like COIL, the gas velocity inside the cavity (U) along with the mirror dimension in the flow direction (l_{res}) determines the residence time ($t_g = l_{res}/U$) of the gas in the gain medium and hence the utilization of the medium for complete power extraction. Therefore, in COIL system, the power extraction efficiency (η_{ext}) is governed by the product of the extraction efficiency, for extracting power from the gain medium (η_{extm}), and the extraction efficiency of the resonator (η_{extr}):

$$\eta_{ext} = \eta_{extm}\eta_{extr} \qquad (26)$$

In addition, the extraction efficiency is also dependent on effective mixing of lasing specie (iodine) with pumping gas flow in the laser cavity (to be discussed in the next section). The resonator extraction efficiency is dependent upon the type of the optical resonator, the mirror losses and is expressed as,

$$\eta_{extr} = \frac{t}{t+a} = \frac{g_{th}2L-a}{g_{th}2L} \qquad (27)$$

Where t and a are the transmission and absorption loss coefficients of the resonator mirrors respectively, L is the gain medium length and g_{th} is the threshold gain of the resonator. For a resonator with output coupler and rear mirror reflectance as R_{out} and R_{max} and no other losses, the threshold gain can be expressed as,

$$g_{th} = \frac{\ln(R_{out}R_{max})}{2L} \qquad (28)$$

For the resonators with other losses like scattering (S_{out}) and diffraction (δ) the resonator extraction can be easily estimated using the relation [27],

$$\eta_{extr} = \left(1-R_{out}-S_{out}\right)\Big/\left[\left(1-R_{out}-S_{out}\right)(1+\delta)+S_{out}+\left(R_{out}/R_{max}\right)^{\frac{1}{2}}\left(1-R_{max}\right)\right] \qquad (29)$$

On the other hand, the medium extraction efficiency describes the availability of the laser medium that depends on the flow conditions of the pumping and lasing mediums. Normally for the gas flow lasers, the Rigrod relation can express the medium extraction efficiency as,

$$\eta_{extm} = 1 - \frac{g_{th}}{g_o} \qquad (30)$$

Where g_o is the small signal. Usually, Rigrod relation can be applied in laser systems where the lasing species is same as the energy containing species thereby allowing direct power extraction as the gain saturates. However, in a COIL device it is slightly different because the energy initially transfers from O_2 ($^1\Delta$) to the iodine lasing species and the power is extracted from excited iodine as the gain saturates. Therefore, the COIL medium extraction efficiency and the saturation intensity can be expressed using by modified Rigrod's relation

[22, 28]. Barmashenko et al[28] have shown that in case of fast relaxation where the relaxation time of the upper level τ is far less than the residence time t_g ($\tau << l_{res}$), the gain in the medium gets homogeneously saturated and Rigrod relations are valid for any kind of resonator. However, for slow relaxation processes where $\tau >> l_{res}$ as is the case in most supersonic COILs, the medium extraction efficiency for a simple resonator can be written in the form;

$$\eta_{extm} = \left(1 - \frac{g_{th}}{g_o}\right) \Bigg/ \left(1 - \Phi \frac{g_{th}}{g_o}\right) \qquad (31)$$

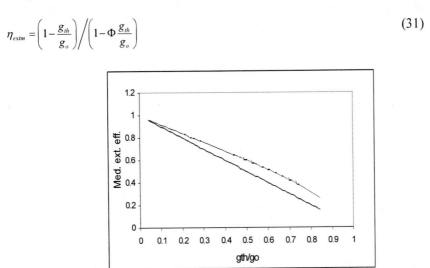

Figure 6. Typical medium extraction efficiency variation with g_{th}/g_o by both Rigrod (continuous line) and Barmashenko (dotted line) model [28].

Where ϕ is a COIL chemistry parameter that varies between 0 and 1 and is given as a function of $Y_{\Delta,cavity}$ as

$$\Phi = \left[Y_{\Delta cavity} - \frac{1}{2K_e + 1}\right] \Bigg/ \left[Y_{\Delta cavity} + \frac{1}{K_e - 1}\right] \qquad (32)$$

For a typical condition of $g_{th}/g_o = 0.25$, $Y_{\Delta,\,cavity} = 0.45$ and $T = 300^o K$ the chemistry parameter is ~0.43. The typical variation of medium extraction efficiency for different values of g_{th}/g_o by both Rigrod and Barmashenko model are shown Fig.(6). For this case, the medium extraction efficiency using Rigrod relation and relation (32) comes out to be 0.75 and ~0.84 respectively.

2.6. Mixing Efficiency

In most COIL devices, the iodine along with its carrier gas (He or N_2), is injected transverse into the primary flow, consisting of $O_2(^1\Delta)$, $O_2(^3\Sigma)$, unutilized Cl_2, H_2O vapor and the buffer gases. After the injection, the secondary jets will change the shape and direction (bend towards the primary flow direction) and merge after traveling a short distance from the injection plane and thus mixing is achieved at an optimum distance from the mixing plane.

Cohen et al. [29] have arrived at the condition which suggests that the penetration of two cross flow jets is proportional to the momentum ratios of the two streams $[\rho_s V_s^2 / \rho_p V_p^2]$, where ρ and V represents the effective density and velocity of the primary and secondary flows suffixed as p and s respectively. Hence the control of secondary or even primary flow parameters can produce optimal penetration conditions and therefore proper mixing. The effect of secondary gas flows on the penetration parameter and hence the laser power has been investigated by many workers [23, 30] Quantitatively, the penetration of the secondary flow into the main flow is represented by the relative penetration parameter, $\Pi_{relative}$ [30], which is the measure of COIL mixing efficiency and is represented as,

$$\Pi_{relative} = \frac{\Pi_{flow}}{\Pi_{full}} \qquad (33)$$

where, Π_{flow} corresponds to the relative gas momentum of the two streams whereas Π_{full} is a function of the geometrical parameters of the flow ducts.

$$\Pi_{flow} = \left(\frac{n_{sec}}{n_{pri}}\right)\left[\frac{W_{sec} T_{sec} P_{pri}}{W_{pri} T_{pri} P_{sec}}\right]^{0.5} \qquad (34)$$

where, subscripts 'pri' represents all the species in the main flow: oxygen, primary diluent nitrogen and unutilized chlorine, 'sec' represents all the species in the secondary flow: iodine and secondary nitrogen. Also, W and n represent the corresponding average molecular weight and molar flow rate respectively; T and P are the temperature and pressure in the corresponding flow line respectively.

The geometrical penetration parameter (Π_{full}) is evaluated from the hardware data as,

$$\Pi_{full} = \frac{dA_s}{5DA_p} \qquad (35)$$

where, d is the height of primary flow channel, A is the flow cross-section; D is the diameter of the secondary flow orifice. Usually, relative penetration parameter in the vicinity of unity is an indicative of proper mixing ($\eta_{mix}=1$) in COIL systems. When η_{mix} is less than unity, it affects the extraction efficiency indirectly as the gain depends on $Y_{\Delta,cavity}$ and T_{cavity}. $Y_{\Delta,cavity}$ is affected because of the change in Y_{diss} as in case of $\eta_{mix}=1$, the dissociation yield , given in eqn (23), will get modified as

$$Y_{diss} = \frac{N \overset{o}{I_2} F}{M_{Cl2} \overset{o}{U}_{Cl} \eta_{mix}} \qquad (36)$$

Further, additional heat released during this improper mixing will result in higher plenum and cavity temperatures in an inhomogeneous manner thus reducing the gain across the flow. Bruins et al. [31] have suggested that the value of η_{mix} can be estimated from the gain (g) values across the flow direction using relation

$$\eta_{mix} = \frac{\int_0^H g.dy}{g_{max} H} \tag{37}$$

Where H is the height of the flow duct across (y) the flow direction (x) and g_{max} is the maximum value of the gain obtained.

3.0. CHEMICAL-OXYGEN IODINE LASER: SUB-SYSTEMS

The schematic of a COIL system is given in Fig. (7). In a typical COIL system, the pumping medium (singlet oxygen) is generated in a chemical reaction chamber viz. Singlet Oxygen Generator (SOG). The generated singlet oxygen in SOG is mixed with primary buffer gas and is transported through a transport duct towards the laser cavity for COIL operation. In between, the lasing medium, iodine, in vapor form along with the secondary buffer gas is also injected which mixes with the singlet oxygen flow. The mixed flow, consisting of singlet oxygen, buffer gas, iodine molecules / atoms, is allowed to expand adiabatically into the laser cavity to create favorable lasing conditions. A steady flow of the gain medium is required to be maintained (in a CW COIL) employing continuous generation and supply of pumping and lasing medium into the laser cavity which is connected either to a large capacity evacuation system or a pressure recovery system. The unutilized chlorine from the SOG and the used iodine are trapped in a liquid nitrogen trap situated in between cavity and the evacuation system.

From the realization point of view, COIL is a complicated engineering system, which includes many subsystems as in evident from the schematic shown in Fig. (7). The major subsystem includes: the singlet oxygen generator, the iodine supply system, Supersonic nozzle assembly, optical resonator, cold trap and the evacuation system/pressure recovery system. This section highlights the major developments in these critical areas.

3.1. Singlet Oxygen Generators

Since the invention of metastable singlet oxygen, $O_2 (^1\Delta_g)$, its applications have been well known in various fields like Biological [32], Photochemical [33] and Physical Chemistry [34]. R. G. Derwent [35] was the first to suggest the use of singlet oxygen as the pumping source for Iodine lasers. The idea has been exploited because of following features of singlet oxygen:

1. The energy level of $O_2(^1\Delta_g)$ state is very near to the energy level of iodine atoms and hence results in near resonant energy transfer between them leading to high efficiency lasing. This can be seen from the corresponding energy level diagrams.

2. Further, the large radiative lifetime (45 minutes) of $O_2 (^1\Delta_g)$ allows it $\sim 10^6$ wall collisions without deactivation. Therefore it is possible to generate and transport $O_2(^1\Delta_g)$ without significant losses.

3.   Further, in COIL the fraction of O_2 $(^1\Delta_g)$ in the flow controls the population inversion in iodine atoms and hence the small signal gain. Thus, power scaling up is a direct function of the singlet oxygen generation.

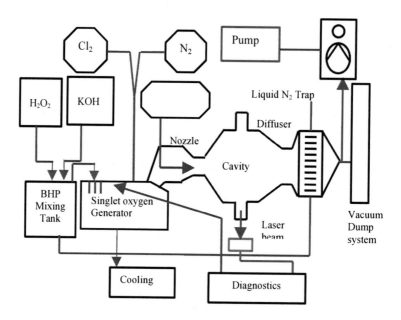

Figure 7. The schematic of the COIL System.

Initial experiments for generation of singlet oxygen were confined to small quantities required for photochemistry applications. However the COIL applications where a huge quantity with large fraction of singlet oxygen (> 15 %) is required demand special techniques for its generation. Therefore, most critical in the engineering of COIL is the development of the SOG and hence it is considered to be the heart of the COIL system. The most commonly used generators fall in the chemical category. Following section briefly discusses these generators with special focus on jet generator, whose technology has matured to an extent that it is being used in all the present day COIL based systems.

The classical method to produce the singlet oxygen in laboratory has been through the reaction of hydrogen peroxide and chlorine or bromine containing compounds like sodium hypo-chlorate, hypo-bromates, b-butyl hypo-chlorite ClO^-, Organic chlorinated compounds [36-39]. For practical COIL applications, the techniques have been refined and even today the most advanced COIL systems are based on generation of singlet oxygen through chemical means. Most commonly used reaction involves the gaseous Cl_2 and aqueous solution of H_2O_2 mixed with a strong base, MOH (M=Na or K). The hydroxide ions react with excess of H_2O_2 as,

$$HO^- + H_2O_2 \Leftrightarrow HO_2^- + H_2O \qquad\qquad (38)$$

The equilibrium of the above reaction is strongly shifted towards right, hence practically only HO_2^- ions will exists in Basic Hydrogen Peroxide (BHP). These HO_2^- ions react with the chlorine gas molecules.

The reaction occurs in the liquid phase between the trapped chlorine and HO_2^- ions, which are produced in the basic solution and given by the most widely accepted set of chemical reactions [21];

$$H_2O_2 + MOH \rightarrow M^+ + HO_2^- + H_2O \qquad (39)$$

$$HO_2^- + Cl_2 \rightarrow O_2(^1\Delta) + 2Cl^- + H^+ \quad k = 2.7 \times 10^{10} \ cm^3/mol.sec \qquad (40)$$

$$2M^+ + Cl^- \rightarrow 2MCl \qquad (41)$$

Singlet oxygen production as per reaction (eqn. 40) is 100% (with respect to the utilized chlorine) within the surface of BHP liquid column. The significant reduction in experimentally observed singlet oxygen outlet is mainly due to the loss of singlet oxygen within the liquid column itself. The major mechanism of this two-phase reaction can be explained from the following schematic Fig. (8). While both mediums are allowed to interact, the chlorine molecules can be either absorbed or reflected from the interface. The absorbed chlorine molecules can diffuse into the liquid and react with the hydroxyl ions present in the BHP liquid medium and thus generate the singlet oxygen molecules. The generated singlet oxygen molecules inside the liquid surface can diffuse out of it without loosing its energy and may be extracted out of the generator. It can also happen that, inside the liquid medium that the generated singlet oxygen molecules can quench into ground state oxygen molecules while diffusing out. Further, once even the singlet oxygen molecules are detached out of the liquid column, before their extraction out of the generator volume it can again be re-absorbed many times into the liquid column and can loose its energy thus being converted into ground state oxygen molecules. In addition to all these liquid quenching mechanism, during the extraction out of the generator volume, the singlet oxygen molecules can also undergo self pooling and other gas phase quenching with chlorine, water vapor etc. Due to these complexities, it is impossible to extract 100 % of singlet oxygen molecules out of the generator.

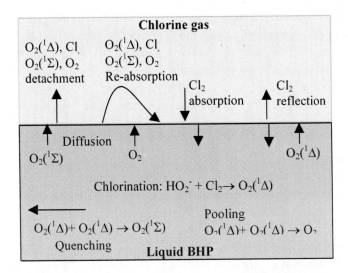

Figure 8. Schematic of the kinetics two-phase reaction between chlorine gas and liquid BHP.

The chemical generator, which can yield maximum fraction of singlet oxygen molecules out of it, needs to minimize all the above said problems. The basic requirement for minimum liquid phase quenching is that the chlorine diffusion depth (singlet oxygen generation depth) in the liquid has to be maintained sufficiently small so that the singlet oxygen diffusion time out of it will also be small and insufficient for the deactivation to take place. In other words, the O2 ($^1\Delta_g$) residence time inside the liquid column should be much less than the liquid phase deactivation rate. The minimum Cl_2 diffusion length (l_{cl}) required for the O2 ($^1\Delta_g$) generation can be calculated from the diffusion constant (and the residence time substituted with reaction rate, k, for reaction (40) as

$$l_{cl} = \sqrt{2D_{cl}\tau_{cl}} = \sqrt{2D_{cl}(k[HO_2^-])^{-1}} \qquad (42)$$

Typically, for $D_{cl} = 7\times10^{-6}$ cm^2/sec, $k = 2.7\times10^{10}$ cm^3/mole-sec, $[HO_2^-] = 2$ Moles/liter, the value of l_{cl} will be about 5×10^{-7} cm. Therefore, the residence time of singlet oxygen for diffusion out of it can be calculated for the oxygen diffusion coefficient same as chlorine as,

$$\tau_o = \frac{l_{cl}^2}{2D_o} = \frac{\left(5\times10^{-7}\right)^2}{2\times7\times10^{-6}} \approx 2\times10^{-8} \text{ sec} \qquad (43)$$

Here it is worth noting that this required residence time is much less than the liquid deactivation rate ($\tau_{\Delta l} = 10^{-5}$ sec) of singlet oxygen in liquid medium;

$$O_2\,(^1\Delta_g) \rightarrow O_2\,(^3\Sigma_g) \text{ (liquid)} \qquad\qquad k = 5\pm2\times10^5 \text{ sec}^{-1} \qquad (44)$$

This demands that for an efficient singlet oxygen extraction mechanism, the two-phase reaction should occur at the surface preferably within a very thin layer (~100 A°). The gas phase quenching can also be minimized utilizing optimum operation parameters such as pressure, the gas residence time etc.

The main requirements of SOGs from the viewpoint of COIL operation are the high chlorine utilization (more than 90%) and the high singlet oxygen yields (more than 60%). This requirement has resulted in a vigorous research in COIL field to produce the chemical generators with suitable geometries and techniques to achieve the desired performance. These include bubbler, rotating disk, jet type, aerosol, mist type etc. [24,40-75]. Brief description of some of the important chemical singlet oxygen generators is given in the following paragraphs.

3.1.1. Bubbler SOG

Bubbler SOG is the first of its kind to deliver a singlet oxygen yield sufficient to demonstrate COIL power. The method uses a bottle with BHP filled to certain level where the chlorine gas is bubbled in the solution, reacting and producing singlet oxygen on its way up through the liquid. The schematic of the bubbler type SOG is shown in Fig. (9).

The first experiment [2] itself has demonstrated a singlet oxygen yield of about 35%. Even though the initial power demonstration was about few milliwatts with this SOG, an immediate improvement resulted in the [41] demonstration of 100W output power. Spalek et al. [63] have reported systematic studies on the optimization of various conditions such as depth at which the chlorine is bubbled to achieve a yield greater than 50%. Initial COIL power extraction experiments have been conducted using this type of singlet oxygen generator. In 1987, Yoshida [64] and his team has used a modified bubbler SOG with dense array of bubbler tube close to the BHP surface and reported a significant improvement in the singlet oxygen yield and obtained about 200W power. Later, Shimizu et al. [76] scaled up this type of generator to obtain a power of 1kW. The Air force Research Laboratory (AFRL) [42] has used this technique to achieve power levels up to 4.6kW.

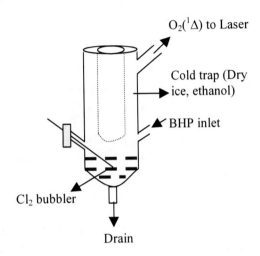

Figure 9. Schematic of the bubbler SOG.

The major disadvantage of this technique is that the singlet oxygen yield is relatively less, especially when large quantity is required at high pressures. It follows from the fact that the pressure of the gas in the bubble (chlorine + singlet oxygen) is several times higher compared to that of the gas above the liquid column, due to the additional hydrostatic pressure of the liquid [63] column on the bubble. This results in the dimolar pooling of singlet oxygen resulting in smaller yields above the liquid column. For example, if the bubble is at 10 cm below the liquid surface, then the pressure on the bubble ($h\rho g$) will be about 9.5torr higher in addition to the pressure above the liquid column. It means that if there is a 3 torr operating pressure above the liquid column, the bubble pressure will be about 12.5torr. Spalek et al. [77] has carried out a detailed theoretical investigation by considering the volumetric fraction of gas in gas-liquid mixture.

The typical operating pressure of this bubbler is few torr and hence is suitable only for subsonic COIL operation. One of the other critical issues in low operating pressure system is the large water vapor fraction (ratio of water vapor pressure to the oxygen pressure). This large water vapor fraction requires efficient water vapor trap in between the SOG and laser cavity, which in turn results in large transport duct.

3.1.2. Rotating Disc SOG

In order to overcome the issues of low pressure, low velocity, large duct volume and large water vapor fraction, the research focus was shifted towards alternative ideas, which can yield large throughput (high pressure and velocity) generators, suitable for efficient supersonic operation of COIL. The development of the rotating disc generator is the major breakthrough in COIL research. The idea is to provide maximum reaction surface between the BHP solution and chlorine gas with replenishment of fresh liquid film for the gaseous chlorine. Here a series of rotating discs, approximately half dipped in solution, generate a fine BHP film over their surfaces for efficient reaction with chlorine gas. In conventional rotating disc generators the shaft of the disc pack is led through the generator housing and connected to an electric motor. Normally the speed of these rotors is about 20-30 rpm. The efficient operation demands fast replenishment of BHP, which in turn requires larger speeds of rotation of the discs. But in practice, once rotation speed exceeds 30-40 rpm, BHP entrainment starts leading to BHP carry over. Typically, a kilowatt level laser would require ten nos. of 13" – 15" diameter disks with 2.5 – 3 mm spacing [78]. This type of generator has been able to produce a COIL power of about 40 kW [42] when used in a modular fashion as shown in Fig. (11).

Though not much work on up scaling of power beyond 40 kW has been reported using rotating disk SOG, however, its smaller version known as RADICL has been used for conducting various tests for basic understanding of COIL. These tests include 2-D gain maps, iodine dissociation studies, magnetic gain switching, mode locking, metal cutting demonstration etc. A detailed review of the major results has been given by Helms [79]. Schall and Duschek [67] demonstrated a new variant of rotating disc generator (called as "Auto disc rotating generator") without electrical motor. In this case the momentum of the continuously flowing gas controls the rotations of the discs. The authors have been able to use this type of generator for COIL operation resulting in few hundred watts of power. Yang et al. [80] has used a simpler structure of rotating mesh in place of rotating disk. The technique has been adapted for making the system lightweight as well as for enhancing the partial pressure of singlet oxygen. In a system that was designed for a 5 kW level, it was possible to achieve 3.7 kW power with partial pressure of oxygen as 10.6torr as compared to 8torr achieved in rotating disk experiments.

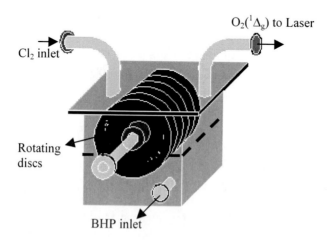

Figure 10. Schematic of the Rotating disc generator.

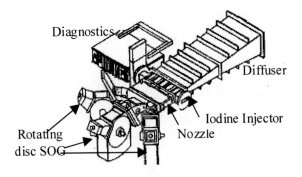

Figure 11. Modular 39 kW rotating disk based COIL system [42].

The major limitation of rotating disk SOG based COIL system is that one cannot produce large quantities of singlet oxygen under high partial pressures. Under best operating conditions, the oxygen operating partial pressure in the generator is usually never more than 10 – 12torr and the singlet oxygen yield is 40 – 50 %.

3.1.3. Jet Singlet Oxygen Generator

The search for chemical singlet oxygen generators with higher specific surface area and high throughput led to the invention of jet type singlet oxygen generator (JSOG) by Balan et al. [72]. The jet type SOG is the most advanced singlet oxygen generator, which has proved its potential over the other techniques even for high power COILs. The technology of this type of generator has matured very rapidly during the last decade and most of the important issues have been well understood. The main driving force behind the rapid progress is its potential of scaling up which is evident from the fact that most present day advanced COILs including Airborne Laser are based on jet SOG. Keeping in view the importance of this generator, we have tried to discuss this generator in great depth. In this technique, the working solution is supplied into the reaction chamber in the form of large number of fine jets, where the chlorine gas reacts at the periphery of the jets. This technique further makes the replenishment of BHP faster even at high pressure. The scheme of the first JSOG as suggested by Balan et al. [72], is as shown in Fig. (12).

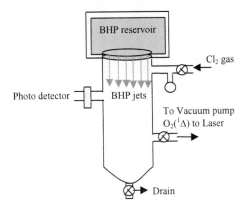

Figure 12. Schematic of the first jet type SOG.

The jet singlet oxygen generator has the following advantages over the other chemical generators:

1. In jet generators, the droplets generated at the BHP injection holes travel almost in the same direction of the jets and suitable selection of jet geometry can makes it possible for self-separation of the drops.
2. Since the jet stability depends on the relative momentum between the gas and jets, that makes it possible to operate JSOG with relatively high gas velocities by adjusting the jet flow parameters,
3. Fast Cl_2 absorption due to high specific surface (controlled by jet diameter and distance between jets) of the generator.
4. High gas velocity operation allows proportional increase of jet velocity and hence fast BHP surface renovation (due to high jet velocity and hydrodynamic effects).
5. Efficient liquid-gas heat exchange eliminates gas heating in quenching reactions
6. Water vapor partial pressure is close to the saturated one for initial BHP temperature of $< -10°C$

However, the operation of JSOGs utilizing above advantages need proper engineering and optimization of their parameters. For example, the droplets are generated at the jet injection holes and the size and number of droplets generated depend on the jet injection hole geometry, quality of the hole and liquid flow parameters. Further, jet stability depends on the liquid flow parameters, jet properties and the gas flow parameters relative to the liquid flow. Gas heating inside the generator highly depends on the gas flow parameters like pressure and residence time. Fast BHP replenishment depends primarily on the BHP supply scheme.

The required chlorine flow rate through a SOG for a desired COIL output power depends on the chemical efficiency (η_{chem}) of the entire system, which strongly depends on the SOG output performance as discussed earlier. For an expected chemical efficiency this requirement of gaseous fuel flow rate is estimated using the relation [22],

$$M_{Cl_2} = \frac{P_{out}}{91\eta_{chem}} \qquad \text{mmol/s} \qquad (45)$$

For example, for a COIL output power of P_{out}=500 Watts and considering practically achievable η_{chem} =0.2, the required chlorine flow rate ~ 27.5mmole/s.

The performance of the JSOG depends on the optimization of various parameters like the gas pressure, velocity, generator geometrical parameters, BHP composition and temperature etc. The generator pressure determines the gas concentration in the active region that helps in estimating the flux of the reaction species in the generator. The generator pressure along with the area of cross-section of the generator (A_{SOG}) and the gas flow rate into the generator (M_{Cl}) determines the gas velocity (U_g) in the reaction zone. These parameters are related as,

$$A_{SOG} = \frac{M_{Cl} RT}{P_{SOG} U_g} \qquad (46)$$

Further, the chlorine utilization and singlet oxygen yield are strong function of jet length. Practically, it is not possible to have larger jet length because of their tendency for early break up. This jet break up can lead to the formation of droplets resulting in carryover into the cavity. Therefore, the generator length or the jet length is usually decided from the viewpoint of stability, which depends on the following factors:

- Fluid properties (Parameters like Reynolds number, Weber number etc.).
- Driving force above the liquid column.
- Geometry of the jets (Shape, diameter etc.).

The jet break up length can be estimated using the generalized formula given by Weber for the case of cylindrical jets [81].

$$L/d = C\ W_e^a\ F(oh) \tag{47}$$

where, C is the size of the initial disturbance and assumes value between 1 to 20, 'a' is a constant and is considered as 0.5 for cylindrical jets and $F(Oh)$ is the modification factor taking into account of dissipation effects (ranges between 1 to 1.5). The Weber number can be defined as,

$$W_e = (\rho v^2 d) / \sigma \tag{48}$$

where, ρ is the density of the liquid, v is the velocity of its flow, d is the diameter of the jet and σ is the surface tension of the liquid. Therefore, for an extreme case corresponding to C =1 and $F(Oh) = 1$, jet breakup length can be written as,

$$L_{minimum} = \rho^{0.5}\ v\ d^{1.5} / \sigma^{0.5} \tag{49}$$

In most case of BHP solution, $\rho = 1300$ Kg/m^3 and for a typical value of $d = 0.8$ mm, $v = 6$ ms^{-1} and $\sigma = 0.074$ Nm^{-1}, one gets $L_{minimum} = 18$ cm. In practical situation, this value is expected to be less because of the presence of the aerodynamic forces due to the counter flow gas stream conditions inside the reaction zone. Therefore, for a practical situation, a jet length of 10 -15cm is ideal.

Zagidullin et.al [73,75,82] have suggested a theoretical model regarding kinetics of a JSOG based on their experimental results. Based on this model, in order to attain a chlorine utilization of more than 90% and singlet oxygen yield >60%, the design parameters of the JSOG are to be selected such that the following three conditions are satisfied.

(i) The 'B' parameter:

$$B = \frac{2[Cl_2]_o}{[HO^-_2]\sqrt{\pi D_H}} \sqrt{\frac{\beta U_g}{\sigma U_l}} \leq 1 \tag{50}$$

(ii) The reduced length of the active zone of the reaction vessel,

$$l = \frac{\beta \sigma L}{U_g} > 3 \qquad\qquad (51)$$

(iii) An 'A' parameter describing the net singlet oxygen quenching inside the reactor as,

$$A = \frac{k_{abs}[Cl_2]_0}{\beta \sigma} \leq 0.1 \qquad\qquad (52)$$

Where $[Cl_2]_0$ is the concentration of chlorine molecules at the admission point, $[HO_2^-]$ is the concentration of HO_2^- ions in the jets, D_H is the diffusion coefficient of HO_2^- ions ($\approx 1.5 \times 10^{-5}$ cm^2/sec), σ is the specific surface area, U_g and U_l are the gas and liquid velocities at the reaction zone, L is the jet length or the generator length and β is the surface reaction rate expressed in terms of the rate constant of the reaction:

$$Cl_2 + HO_2^- \rightarrow 2Cl^- + H^+ + O_2(^1\Delta) + 27 Kcal/mol \quad k \sim 4 \times 10^8 \; lr/mol/sec \qquad (53)$$

and is estimated using the relation

$$\beta = \frac{\sqrt{K[HO_2^-]D_{Cl}}}{m + \left(\dfrac{4}{v\gamma}\right)\sqrt{K[HO_2^-]D_{Cl}}} \qquad\qquad (54)$$

In the above relation D_{Cl} ($= 2 \times 10^{-6}$ cm^2/sec) is the diffusion coefficient of Chlorine in the solution, m is the Henry constant (≈ 0.3), v is the thermal velocity of the chlorine molecules, and γ is the probability of trapping of the chlorine molecules at the surface of the solution ($\gamma \approx 1$). k_{abs} represents the $O_2(^1\Delta)$ quenching rate constant $\approx 10^{-16}$ cm^3/sec For example, for a gas velocity of 15 ms^{-1}, $[HO_2^-]$ of 6.5 mol/l, a jet length of 10 cm, specific surface area of 5 cm^{-1} and a liquid velocity of 6 ms^{-1} the value of B, l and A turnout to be ~ 1, ~ 7.7 and 0.1 respectively ensuring more than 90% chlorine utilization and 60% yield.

Barmashenko et.al has proposed a general theoretical model [83,84] for the chemical generators suitable for almost all kind of generators viz. jet type generators, bubbler column, film, aerosol etc. The model can be applied to different kind of geometries such as co-flow, counter flow conditions of gas and liquid streams in the jet SOG. The model is capable of predicting the expected singlet oxygen yield and utilization by accounting for the singlet oxygen quenching in the liquid phase (inside the BHP jets) by defining an analytical relation for characteristic pressure, which corresponds to zero quenching condition. This model is extremely helpful in predicting the performance of the SOG under different operating conditions.

According to this model, for a case of negligible gas phase quenching inside the generator with flows of BHP jets and Cl_2 being anti-parallel, the analytical relations for chlorine utilization and singlet oxygen yield at the exit of the generator can be expressed as

$$\text{For,} \quad P > P_{OL}\sqrt{\frac{2l_g \sigma D_{Cl}}{U_g K_{Cl}\sqrt{D_{O2}}\,\tau}} \qquad\qquad (55)$$

$$U_{Cl} = \frac{P_{OL}}{P} \sqrt{\frac{2l_g \sigma D_{Cl}}{U_g K_{Cl}\sqrt{D_{O_2}}\tau}} \tag{56}$$

$$Y = \frac{1}{U_{Cl}} \frac{1}{\sqrt{1-\frac{1}{4}\left(\frac{P}{P_{OL}}\right)^2 (1-U_{Cl})}} \left\{ \tan^{-1}\left[\frac{\frac{1}{2}(1+U_{Cl})-\frac{P}{P_{OL}}}{\sqrt{1-\frac{1}{4}\left(\frac{P}{P_{OL}}\right)^2 (1-U_{Cl})}}\right] - \tan^{-1}\left[\frac{\frac{1}{2}(1-U_{Cl})-\frac{P}{P_{OL}}}{\sqrt{1-\frac{1}{4}\left(\frac{P}{P_{OL}}\right)^2 (1-U_{Cl})}}\right] \right\} \tag{57}$$

For, $\quad P < P_{OL}\sqrt{\dfrac{2l_g \sigma D_{Cl}}{U_g K_{Cl}\sqrt{D_{O2}}\ \tau}}$ $\tag{58}$

$U_{Cl} = 1$ and

$$Y = \frac{P_{OL}}{P}\tan^{-1}\left(\frac{P}{P_{OL}}\right) \tag{59}$$

Where, P is the partial pressure of chlorine and P_{OL} is the characteristic generator pressure for which zero liquid phase quenching of singlet oxygen occurs and is given by,

$$P_{OL} = \sqrt{\frac{D_H K_{Cl}\sigma\sqrt{D_{O2}}\tau}{2\pi D_{Cl}}}[HO^-_2](kT)\sqrt{\frac{U_l}{U_g}} \tag{60}$$

In addition to all above parameter optimization, the selection of configuration and geometry of the JSOG should be proper to get a stable operation. The major criteria with the selection of suitable geometry are the stable and smooth operation of this jet generator. The criticality of the two-phase interaction is that the gas flow should not perturb the jets and the hydrodynamic stability should be achieved in the generator. The catastrophic carryover of the liquid to the laser may occur and impair lasing. In addition to this major liquid carryover, a secondary carryover is also possible in JSOGs. This secondary carryover consists of minute droplets being carried along with the exit flow, which are generated at the liquid jet injection points. These droplets tend to travel along the liquid jet direction and can be carried over along the gas flow as soon as their momentum becomes less compared to the gas momentum.

Various kinds of geometries of jet SOG have been attempted in order to have better efficiency and high throughput (pressure and velocity) generation. The SOGs can be operated in various configurations depending on the approach in which the gaseous (Cl_2) and the liquid fuel (BHP) are fed [82].

Based on the chlorine feeding direction with respect to the liquid jets, there are three possible designs of jet SOG: (a) Co-flowing, (b) Cross-flowing and (c) Counter flowing as shown in Fig.(13). The major criterion with the selection of suitable geometry is the stable and smooth operation of the jet generator by ensuring hydrodynamic stability and negating the possibility of catastrophic carryover along with minimizing the secondary carryover as far as possible.

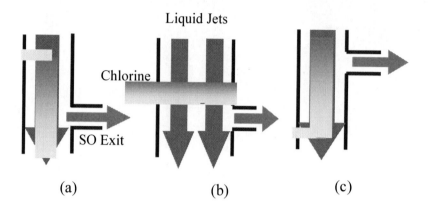

(a)   (b)   (c)

Figure 13. Various employed geometries of JSOG.

Detailed studies [4] have shown that the counter flow geometry is the most suitable for COIL application because of reliability and its ease of operation. This is mainly due to the fact that gas needs relatively high momentum to perturb the liquid column thereby making it possible to operate at high gas velocity with minimum $O_2(^1\Delta)$ quenching. Further, the efficiency of counter flow geometry based COIL is significantly higher as compared to cross flow one. Fig (14) shows typical cross flow geometry used by McDermott et al. [4], where its performance has been compared with that of a counter flow geometry. It has been observed that under similar conditions, the efficiency of cross flow generator was 11.9 % as compared to 18.2 % achieved for counter flow conditions. Based on the success of counter flow geometry, the authors not only up scaled the system to more than 10 kW level but also increased the efficiency up to 29.6 % by operating the system under optimized conditions.

However, where very high powers are required and the system need to be compact like Airborne Laser (ABL), the counter flow configurations prove to be bulky, as one requires coupling of many modular SOGs. In such cases, the cross flow configuration is an obvious choice. The injection of chlorine and the other operating conditions, nevertheless, become very critical to avoid instabilities in the system.

The catastrophic and secondary carryover of the jet generators can also be controlled by the way the generated singlet oxygen is extracted from the generator. Initial studies have used the horizontal type of system [47] for operation of COIL, which sometimes results in the carry over of droplets of BHP into the cavity. Recently, verti-flow type of system [56,85] has been used to avoid this problem.

The vertical system, however, has been observed to be comparatively much more difficult to handle particularly if one is interested in a smaller system for optimization of specific parameters. Keeping this in mind, a JSOG developed by the authors has been designed and operated with extraction angle as 40°. The system is extremely flexible as a horizontal system, at the same time; the carry over of BHP droplets towards the cavity is minimized similar to a verti-flow system. Figure 15 compares the angular geometry with horizontal and vertical ones. The importance of this generator exit geometry on the stable operation can be easily understood from the following theoretical analysis.

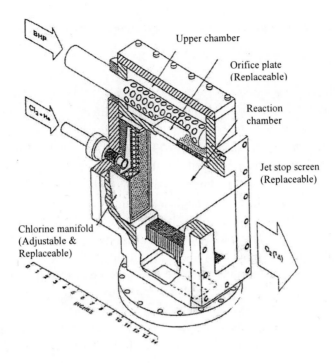

Figure 14. SOG based on cross flow geometry [4].

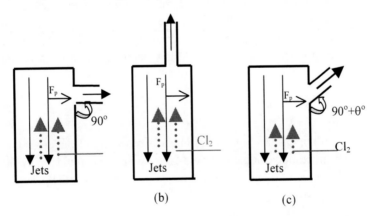

Figure 15. Various geometries of JSOG based on the extraction: (a) Horizontal, (b): Vertical and (c) Angular JSOGs respectively.

The carryover problem in a stable jet generator operation (where there is no jet breakup in the given reaction length) can be divided into mainly two categories:

- Catastrophic carryover of the jets into the outlet.
- Droplets carry over into the outlet.

The first case mainly depends on the hydrodynamic stability, which is a function of the Reynolds number of the jets and aerodynamic force inside the generator. Experimentally it has been observed that the Jet SOG is stable at the pressure up to 100torr and the ratio U_{gas} /

$U_{jet} > 2$[73]. The generator exit geometry has a strong implication on the aerodynamic force acting on the jets as that governs the velocity component in the direction of medium extraction,

$$F_p \alpha\ U_g^2\ Cos^2\theta \tag{61}$$

Where F_p is the force component along the pumping direction, U_g is the magnitude of the gas velocity in the generator and θ is the angle of the generator exit with respect to the horizontal. In the case of a vertical system, $\theta = 90°$, there is no additional drag force acting on the jets in the vicinity of the generator exit thus having the maximum stability. On the other hand, in the horizontal system, $\theta = 0°$ implying maximum drag force on the jets resulting in minimum jet stability. But in the case of angular system with θ value say 40°, the drag forces are reduced to almost half compared to the horizontal system thus assuring better stability thereby having a reduced catastrophic carry over probability.

The second case, the droplets carry over is generally referred to the carry over of the secondary drops, which are generated while the BHP jets are ejected from the BHP injection holes and not because of the jet break up. This drop formation is a function of jet diameter, driving force and injection hole geometry and its finish. The distance of traverse (l) of these drops before coming to rest depends on the initial drop momentum and the opposing drag force, which is present due to the relative velocities of the drop and the gas. If this length l is greater than the generator length (L) then these drops can be separated themselves from the outflow. However, if the $l < L$, these droplets are carried over irrespective of the geometry. For a given generator conditions, the value of drop diameter above which these can be separated out can be estimated in the following manner.

The critical diameter of the drop up to which that can be carried by the aerodynamic force in a given generator length is obtained by considering the energy conservation equation,

$$\frac{1}{2}mU_j^2 - \frac{1}{2}m(U_j - dU_j)^2 = -6\pi\mu rv.dh \tag{62}$$

where, m is the mass of the drop, μ is the coefficient of viscosity of the drop, v is the relative velocity i.e U_g-U_j. The maximum diameter of the drop separable is obtained by integrating the above equation for the velocity range $-U_j$ to 0 within the reaction length 0 to L as,

$$D_{max} = \left\{ \frac{18\mu L}{\rho_l \left[U_l - U_g \ln\left[(U_g + U_l)/U_g \right] \right]} \right\}^{\frac{1}{2}} \tag{63}$$

For example, for a case of L =10 cm, μ =20cp, ρ_l=1300kg/m^3, U_l= 5 ms^{-1} and U_g= 15 ms^{-1}, the minimum separable diameter $D_{max} \approx 182\mu m$. This phenomenon is almost independent of the generator exit geometry or the pumping direction. This occurs because the counter-flow velocities of the jet and the gas govern the drag force on the spherical drops in the reaction zone below the generator exit in all cases.

The geometry of the generator can play a significant role in avoiding carryover of droplets of size greater than 182μm. Due to the presence of aerodynamic forces, the trajectory of the drops moving in the jet direction gets altered and in extreme cases the drops may hit the boundary walls and collapse. This wall wetting is a strong function of SOG geometry and droplet size. The aerodynamic forces direct these collapsed droplets again towards the cavity side. The phenomenon of wall wetting is the most predominant in the droplet carryover. The theoretical aspects dealing with the drop aerodynamics in a JSOG, have been discussed in detail by Rajesh et al. [86] along with a comparative hydrodynamic behavior of an angular geometry with a vertical one. The essence of the study is being briefly presented here.

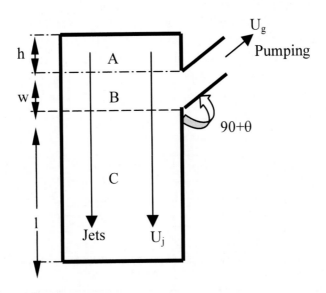

Figure 16. Angular generator configuration.

The angular SOG geometry is shown in Fig (16), three distinct regions of jet flow are identified viz, 'A' is the region where the secondary jet drops are considered to have zero horizontal velocity and typically travel with jet velocity, 'B', the drops are considered to be facing a horizontal component of the aerodynamic force because of the extraction geometry over the length 'w' due to a considered constant velocity component $v = U_g \cos \theta$ (due to small length of the region with $C_D=1$), 'C' is the region where the horizontal traverse is a function of the lift force ($C_L = 0.3$). The well-known equations of motion along with equations for drag and lift force are utilized for determining the overall horizontal traverse.

The corresponding horizontal movement (S_c) in the region 'C' considering an initial velocity gained in region 'B' is expressed as,

$$S_C = \frac{3}{8} \frac{\rho_g}{\rho_l} \frac{l}{r} \left(\frac{U_g}{U_j}\right)^2 \left[wC_D \cos^2 \theta + 0.5 l C_L \right] \tag{64}$$

The horizontal distance traversed by drops of different diameters in region B and C for a case of w=1cm, l= 8 cm, U_g=15 ms^{-1}, U_j= 6 ms^{-1} for all the three geometries (θ =0°, 40° and 90°) are shown in Fig (17) & (18). From the curves it is clear that the horizontal movement in region B in all cases does not exceed 1mm even for 50μm diameter drops. In region C, the drops with diameter less than 50μm can move even few cm distances before these travel over the 10 cm length in both cases with θ =0° and 40° while in case of θ =90° the movement is still about an mm. The estimation shows that in horizontal case (θ =0°) the drops with diameter up to ~ 350μm can move a horizontal distance more than 2 mm and may result in wall wetting. Similar calculations for a vertical (θ =90°) and angular (40°) shows this minimum value as 20 micron and 170 micron respectively implying that the carry over due to wall wetting is insignificant.

Authors have designed and operated a small scale angular jet SOG [Fig. (19)] for more than 600 runs without any catastrophic carryover problem. The design has been based on that of Zagidullin et al. [73,75,82] and Barmashenko et al. [83,84] as discussed above. Power more than 350 W has been obtained with chlorine flow rate of the order of 20 mmol/sec.

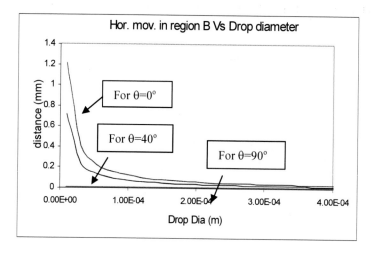

Figure 17. The horizontal movement of the liquid drop.

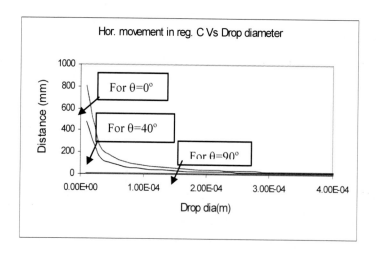

Figure 18. The horizontal movement of the liquid drops in region C for various drop diameters.

Figure 19. Photograph of the angular JSOG.

Many workers [24,69,73,74,75,82,87,88] have reported the results of their parametric studies on jet SOGs. It has been reported that in general, composition of BHP, operating temperature, water vapor contents, gas velocity, liquid velocity, type and geometry of jets, chlorine flow rates, chlorine to primary gas ratio etc. affect the performance of the generator. Some of the representative results are depicted in the following Figs. (20) & (21).

In addition to the most commonly used jet generators, some modified configurations [42,68,71,89] have also been attempted to improve the performance. The considerable types among them are (i) Pipe guided jet generator and (ii) Aerosol or Spray or atomizer type generator. The pipe guided jet generator (moving jets along the surface of pipe arrays) has been reported to improve the stability of the liquid jets inside the generator and the specific surface area [89]. The Aerosol SOGs, also known as the spray type SOG or Atomizer SOG, [42,68,89] are based on the reaction of chlorine gas with an aerosol of BHP prepared using fine spray. These types of SOGs have the advantage of large contact surface (specific surface area) between the chlorine gas and BHP fine drops. However, this technique suffers from the drawback that the fine drops (of BHP) are easily dragged out of the SOG along the singlet oxygen gas flow, which largely degrades the COIL performance. Muto et al. [71] have developed a new type of Mist SOG, that is based on the generation of a fine mist of BHP (drops diameter of the order of about 100μm) using a spray nozzle utilizing either pressurized BHP or a combination of pressurized BHP and compressed gas. Even though the carryover of BHP droplets along with singlet oxygen is reduced to an extent in this method, the amount of water vapor still remains to be too large. Therefore, without using a cold trap, using this technique for COIL applications is not very effective.

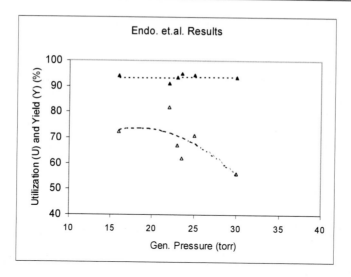

Figure 20. Cl$_2$ Utilization and SO yield variation with generator pressure obtained by Endo et al, for a case of 25 mmol/s of Cl$_2$, Jet velocity was 4.5 ms^{-1} [69].

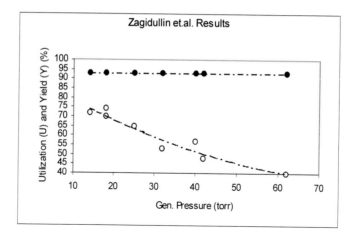

Figure 21. Cl$_2$ Utilization and SO yield variation with generator pressure obtained by Zagidullin et al, for a case of 25 mmol/s of Cl$_2$, Jet velocity was 4.5 ms^{-1}[14].

In order to overcome the droplet carryover, a modified version was introduced, known as Twisted Aerosol, [69, 70] in which the working solution is injected into the reaction chamber normal to the cylinder axis through the holes between the screw blades. Chlorine passes along the helical channels between the blades and interacts with the dispersed solution. In the field of centrifugal forces, oxygen gets free of droplets of the solution, which are thrown to the reactor walls and screw blades. During its rotation, the screw scrapes off the solution from the cylindrical surface of the reactor housing and directs it to a special collector, where, using a centrifugal wheel, the pressure of the spent solution is raised up to atmospheric pressure and the exhausted solution is removed. This type of twisted aerosol has demonstrated high efficiencies and possibility of operating the generator with high pressure and gas velocity. However, this technique requires overcoming many engineering aspects if it is to be used for high power COIL or large quantity production of singlet oxygen.

3.2. Resonators

In order to fully exploit the high power capability of COIL, the resonator design should result in excellent beam quality. The cavity conditions (low pressure and low temperature) of COIL are favorable for realizing very good beam quality, however, the lower gain (0.003—0.004 cm^{-1}) as compared to the other high power lasers like HF/DF, CO_2: GDL, Nd. YAG etc. puts certain restrictions such as using high magnification in unstable resonator geometry. Some of the approaches being followed for power extraction are briefly discussed below:

3.2.1. Stable Resonators

Most COIL systems reported till date have used stable resonators with the demonstrations of high chemical efficiencies. The large aperture in typical COIL system with conventional stable resonator configuration makes them to accommodate many modes and hence the beam divergence and the quality are compromised with the extraction of maximum power. The number of modes generated can be estimated from the Fresnel number and corresponding full angle divergence of the beam are given by the following set of equations (65) &(66) [90].

$$N_F = \frac{a^2}{\lambda L} \tag{65}$$

$$\theta = \frac{\lambda}{\pi \omega_o} \sqrt{\frac{\pi}{2} N_F} \tag{66}$$

Where 'a' is the beam diameter and 'ω_o' is the beam waist and is given by,

$$\omega_o = \sqrt{\frac{\lambda L}{\pi}} \tag{67}$$

Therefore for a typical case with 3 cm aperture radius and mirror separation of 1 m, it can accommodate about 685 modes and hence the divergence will be around 21 mrad.

The typical stable resonator configuration used in COIL systems is a semi-confocal configuration with a combination of a flat (plane) output coupler mirror and a Plano-concave rear mirror with certain separation (L) as shown in Fig.(22).

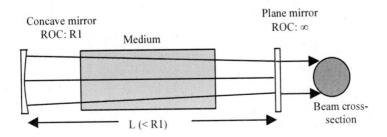

Figure 22. Typical stable resonator configuration for COIL (Semi-confocal).

In a resonator with a combination of flat output mirror and concave reflection mirror, COIL oscillates in high order Hermite Gaussian mode and the divergence angle can be defined as [91],

$$\theta = \frac{a}{\left\{L\left(R-L\right)\right\}^{1/2}} \tag{68}$$

where, 'a' is the beam diameter on the flat mirror, 'R' is the radius of curvature of the concave mirror and 'L' is the mirror separation. One can improve the beam quality to a large extent by adjusting the inter-mirror spacing and choosing the right radius of curvature. However, the ratio of R/L is limited to less than 30 because of the stability of laser oscillation. Taking a practical case of radius of curvature of about 50m and intermirror spacing of 1.6m Fujii and Atsuta [91] have been able to achieve divergence less than 5 mrad for a kilowatt level laser where beam height was of the order of 40mm. The beam height can be reduced without loosing the power extraction by reducing the aperture size employing folded resonator configuration. This can increase the gain length while reducing its cross section and maintain the gain medium volume (beam cross section and gain medium length) as constant. Such configuration has been used in industrial chemical oxygen iodine laser at Kawasaki heavy industries to focus the beam on to a 0.3 mm core optical fiber effectively [92].

The important parameter to be considered for extraction of maximum power is the output-coupling factor, which is given by the relation,

$$t_{opt} = \sqrt{g_o L a} \left[1 - \left(\frac{a}{g_o L}\right)^{1/2}\right] \tag{69}$$

Where, 'g_0' is the small signal gain.

For a given transmittance (t) of the output-coupling mirror, the output power (P_{out}) depends on the saturation intensity (I_s) inside the cavity and its geometry. The saturation intensity for COIL gain medium can be estimated from the relation [24],

$$I_s = \frac{h\nu}{\sigma\tau_{res}\left(\left[I\right]/\left[O_2\right]\right)} \frac{2\left[\left(K_e - 1\right)Y + 1\right]}{2K_e + 1} \text{ W cm-2} \tag{70}$$

Where the terms have their usual meaning and 't_{res}' is the residence time of the gas inside the cavity. Typically the saturation intensity for a supersonic COIL lasing medium is around 2 kW/cm^2 [24]. For, the strongest transition of 3-4 line of the atomic iodine hyperfine structure with dominating Doppler broadening line function, the output power as a function of saturation intensity can be predicted by Rigrod's relation [93].

$$P_{out} = \frac{I_s at}{2}\left[\left\{\frac{2\alpha_{34}L}{t_{opt} + \delta_i}\right\} - 1\right] \tag{71}$$

Where, 'a' is the beam cross-section and δ_i is the resonator internal loss per roundtrip. Various workers while optimizing their lasers have used such resonator configuration [14,41,99]

3.2.2. Unstable Resonator

Since COIL is a low gain laser, unstable resonators are not suitable for low power levels typically up to 10 kW. Even if we take the gain length of 1m, the magnification factor expected for a donut unstable resonator will be of the order of 1.1 and the obtained beam shape would be of no use for practical applications. However, for higher power COIL systems where magnification factor can be more than 1.5, unstable resonator geometry is of great relevance. Most common unstable resonators have two mirrors of different diameters and radii of curvature with larger diameter at the rear side (radius of curvature R_r) and smaller one at the coupler side (radius of curvature R_o). In these kinds of configurations, the output is taken as the diffracted beam around the edges of the output mirror and hence the coupling fractions are larger compared to that in stable resonator case (larger diffraction losses). In such cases, the geometrical output coupling will be given by,

$$\delta = 1 - 1/M^2 \tag{72}$$

Where 'M' is the magnification factor, which is simply R_r/R_o. Such unstable resonators have transverse mode envelopes that can fill large laser volumes but at the same time suppress the higher order transverse modes. The Fresnel number in such resonators are given by,

$$N_F = \left[\frac{(M-1)}{2M^2}\right]\left(\frac{a^2}{\lambda L}\right) \tag{73}$$

Therefore, it clearly indicates that the beam quality can be improved by an order if one can go for a magnification of 4-5. In order to achieve these conditions, certain modified configurations have been attempted. These include conventional confocal unstable resonator and Z pass unstable resonator

Conventional confocal unstable resonators are good for large roundtrip gain medium. This geometry uses a concave mirror of larger radius and convex mirror of smaller radius (both having near 100% reflectivity) and an annular scrap off mirror for output beam diversion. Schematic of such a configuration used by Yang and his co-workers [94,95] is shown in Fig.(23).

With this confocal resonator configuration they could produce an output of 7.1 kW with a laser beam quality of $\beta \leq 6$, where β is defined as, $\beta = \theta_{experiment} / \theta_{theory}$ in which 'θ' is divergence. In order to improve the beam quality further, the authors have demonstrated a novel unstable ring resonator with 90° beam rotation. The magnification used is 1.69 whereas effective cavity length is about 9.82m. The average output power has been of the order of 4 kW with beam divergence as low as 0.27 mrad. The beam rotating resonator not only improves the beam quality but also reduce the influence of the non uniformity on gain distribution inside the medium due to multiple passing [Fig. (24)].

The Z pass configuration makes use of a telescopic unstable resonator geometry employing two plane, one convex and other concave mirrors [96]. This configuration improves the beam quality by reducing the aperture size and by increasing inter-mirror spacing (between the end mirrors) thus increasing the gain medium length. The typical configuration is shown in Fig. (25). With this configuration, for a magnification of 1.3, inter mirror spacing of 3 m and aperture diameter of 3 cm, the beam divergence is ~ 2.1 mrad.

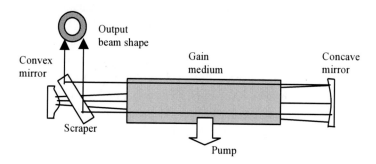

Figure 23. Typical scheme for conventional scheme for confocal unstable resonator.

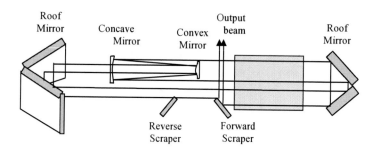

Figure 24. Scheme for unstable ring resonator with 90° beam rotation. [94,95].

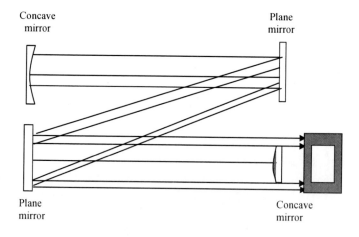

Figure 25. Scheme for Z pass unstable resonator.

3.2.3. Other Resonator Geometries

The emphasis has been to improve the quality of laser beams in COIL systems with power levels in the range of 1 – 10 kW. This is the class of system which has a great potential for material applications. One of the approaches suggested by Yoshida et al. [97] is to employ an intraresonator telescopic . By locating the telescope at a proper position on the optical axis, the resonator can be tailored to have a low beam divergence alomg with improving the resonator stability. The experiments carried out with a magnification factor of M = - 3, indicated that the the multimode divergence could be reduced to 2 – 3 mrad, whereas the divergence for the fundamental mode was even less than a millirad. However, there was a marginal power reduction of the order of 20 – 30 % which was caused by reflection loss on the intraresonator optics.

The other approch in this category has been to employ a stable-unstable resonator geometry suggested by Endo et al. [98]. The resonator is basically a positive branch confocal unstable resonator consisting of a concave and a convex mirror. However, there is a flat part at the centre of the convex mirror. Thus, Hermete Gaussian modes can be established at the core even the medium round trip gain is fairly low. The diffracted out portion of the centre core is further amplified at the outer unstable part and is coupled out from the circumfrence of the convex mirror. As such this resonator results in beam quality as good as central core, while having the capability of oscillating in the extremely low gain medium, where unstable resonators alone are not effective. Detailed theoretical analysis and experimental results have indicated that as compared to stable resonator, this stable-unstable resonator geometry improves the beam quality M^2 factor from 156 to 29. Though the output power is reduced from 35 W to 14 W, but due to the improvement in beam quality, the power density at the focus point improves significantly from 52 to 772 W/cm^2, when focussed using a one meter focal length lens.

3.3. Nozzles and Iodine Injection System

COIL systems in general are capable of being operated in both subsonic as well as supersonic regimes. A comparison of the COIL operation in the two regimes is illustrated in Table-2.

The small signal gain (α) inside the cavity can be expressed as a function of temperature and cavity yield by substituting the relations for threshold yield and equilibrium rate constant in the general gain equation derived earlier.

Table 2. Comparison of COIL operation in subsonic and supersonic regime

	Subsonic COIL	Supersonic COIL
1.	Limited maximum gain, as it is a weak function of pressure	Possibility of obtaining high gain as it is strong function of temperature and increases with decrease in temperature
2.	Highly sensitive to titration ratio (I_2/O_2)	Distinctly less sensitive to titration ratio (I_2/O_2)
3.	Poor power scalability potential as resonator sizes tend to be very large for high flow rate systems	Extremely good potential for power scale up

$$\alpha = 0.5\sigma_o[I]\sqrt{\frac{T_o}{T}}\frac{\{1.5Y_\Delta \exp(402/T)+Y_\Delta -1\}}{\{0.75Y_\Delta \exp(402/T)-Y_\Delta +1\}} \tag{74}$$

It is evident form the above equation that the gain is inversely proportional to the temperature of the flow medium in the cavity [66]. 'σ_0' is the stimulated emission cross-section of the lasing species (iodine atoms), '[I]' is the iodine concentration in the cavity, which is also a direct function of cavity temperature, 'Y_Δ' represents the singlet oxygen concentration inside cavity and 'T' is the cavity temperature. Thus a supersonic expansion resulting in a high Mach number (M) allows in achieving low cavity temperature (T), which can be, expressed as [99].

$$\frac{T_o}{T} = 1 + \frac{(\gamma-1)}{2}M^2 \tag{75}$$

Where, 'γ' is the specific heat ratio of the flow medium and 'T_o' is the stagnation temperature of the cavity flow medium. Thus, if one can increase the cavity Mach number the laser efficiency is expected to improve.

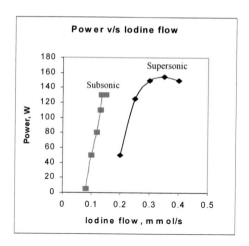

Figure 26. Power variation with Iodine concentration in supersonic and subsonic COILs [9].

Supersonic mode of operating COIL system is preferred primarily because of its inherent advantages in terms of easy scalability to high power regimes. Also, in supersonic operation the optimum power is not highly sensitive to the iodine flow rates [9, 10] whereas in subsonic case it is a very sharp function as can be observed from the Fig. (26).

The conventional nozzles for COIL are usually [I] designed for a cavity pressure of nearly 3-4torr for having the desired small signal gain for optimum lasing. Now from the first principles the cavity Mach number is usually determined using the relation for isentropic flow with the given plenum pressure conditions (typically 25torr in JSOG)

$$P_o = P\left[1 + \frac{\gamma-1}{2}M^2\right]^{\frac{\gamma}{\gamma-1}} \tag{76}$$

Further, it is well known that in a purely contracting flow, the maximum uniform velocity that can be achieved is the sonic velocity. Velocities greater than that of sound can be obtained only by subsequent expansion of stream. This is evident from the following equation as kinetic energy of flow increases at the expense of pressure (i.e. dp is –ive), therefore for M< 1, da should be –ive (i.e. decreasing) whereas when M >1, da should be + ive(i.e. increasing).

$$\frac{da}{a} = \frac{dp}{\rho c^2}\{1 - M^2\} \tag{77}$$

Thus, one ideally requires a converging –diverging passage for achieving supersonic flows. The essential and relevant features of supersonic nozzle design is shown in Fig. (27) below,

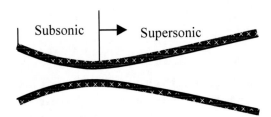

Figure 27. Conventional supersonic nozzle design.

In case of COIL nozzles with the main stream flowing from the plenum and the secondary stream being injected close to the throat (injection mechanism will be discussed later), the region prior to the throat is essentially a straight portion or only marginally converging. The throat dimensions are determined corresponding to the total flow (Main + secondary) to ensure the occurrence of sonic condition at the throat. However, as COIL flows are distinctly complex involving various heat generation mechanisms in the main stream, in mixing phenomenon between the two streams, high temperature of the iodine streams etc. The mixed stream total temperature is determined using the relation corresponding to adiabatic mixing viz,

$$T_{om} = x_p T_{om} + x_s T_{os} \tag{78}$$

Subscripts 'p' and 's' stand for primary and secondary flows and 'm' stands for mixed flow. 'x' represents the mole fraction. Also, in the determination of throat dimensions one has to provide the allowance for the occurrence of the boundary layers thereby diminishing the throat dimensions and providing the possibility of occurrence of a subsonic flow. The boundary layer thickness is known to increase with the extent of heat generation or dumping, thus an estimation of the overall heat being generated in the nozzle needs to be conducted. The relation of boundary layer thickness with the stream Reynolds number is [100],

$$\delta = 5\left(\frac{vx}{U_o}\right)^{1/2} \tag{79}$$

The Reynolds number is defined as the ratio of inertia to viscous force and the dynamic viscosity is a distinct function of temperature which in COIL flows is quiet difficult to predict.

Thus, typically the throat dimensions are determined using the relation for the maximum mass flow rate for the mixed stream parameters of total temperature, characteristic gas constant and specific heat ratio. The mixed stream parameters for the specific heat and the specific heat ratios are,

$$\dot{m}_i = \frac{a_i^*}{\sqrt{T_{oi}}} P_{oi} \sqrt{\frac{\gamma_i}{R_i} \left(\frac{2}{\gamma_i + 1} \right)^{\frac{\gamma_i}{\gamma_i - 1}}} \tag{80}$$

$$C_{pm} = \sum_{i=1}^{n} x_i C_{pi} \tag{81}$$

$$\frac{1}{\gamma_m - 1} = \sum_{i=1}^{n} \frac{x_i}{\gamma_i - 1} \tag{82}$$

The, allowances on the throat height for nozzle are more a matter of experience as the COIL parameters tend to alter from one run to another, however, a value of nearly 15 % is found to be suitable for achieving sonic throat condition in most typical COILs.

The most critical region for design in the nozzle in COIL is the expanding part in order to achieve the conditions of desired Mach number in the cavity. The expanding part is essentially designed by the theory given by Prandtl-Meyer for expansion around a corner, which is more commonly termed as the Method of Characteristics [101]. The basic assumption employed for the design is that the flow is considered to be irrotational and inviscid. The supersonic flow is considered to occur over a curved surface, which can be approximated by an infinite set of straight segments. The change in the local Mach number between any two points is a function of the angle of turn. Thus, the profile generation reduces to that of prediction of Prandtl-Meyer flow. The fundamental equation for the flow –about –a corner is

$$\upsilon = \kappa \tan^{-1} \left(\frac{Cot\alpha}{\kappa} \right) - (90^o - \alpha) \tag{83}$$

Here, υ is the expansion angle or the angle through which the flow is turned in accelerating the flow from local Mach number from unity to the any Mach number M, α is the corresponding Mach angle.

$$\alpha = \sin^{-1} \frac{1}{M} \tag{84}$$

$$\kappa^2 = \frac{\gamma + 1}{\gamma - 1} \tag{85}$$

Thus, if the expansion angle is known in any region, the Mach number can easily be determined. If the subscripts 1 and 2 refer to conditions in regions I and III (refer Fig. (28)), respectively then the angle through which the flow is turned in accelerating the flow from Mach number M_1 to M_2, i.e.,

$$\delta = v_2 - v_1 \tag{86}$$

In other words, the change in the expansion angle is equal to the absolute value of the change in the stream deflection through an expansion region due to a single corner. If, however, the deflection angle is small, then all expansion can be considered to take place along the average Mach line (as shown in Fig. (28)).

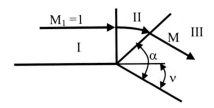

Figure 28. Prandtl - Meyer expansion around a corner.

This line, no longer remains a line of propagation of an infinitesimal disturbance, now it takes on certain characteristic of a shock wave; namely, finite change in direction and Mach number. It is usually referred to as the expansion wave. Thus, the assumption produces a small error, which vanishes in the limit of deflection angle tending to zero. The solution is thus single valued for expansion waves corresponding to weak oblique shocks.

For any given targeted Mach number, while there are infinite number of satisfactory nozzles, there is one in variant parameters; the area ratio of the cavity to the throat.

$$\frac{A}{A^*} = \frac{1}{M}\left[\frac{2}{\gamma+1} + \frac{\gamma-1}{\gamma+1}M^2\right]^{\frac{\gamma+1}{2(\gamma-1)}} \tag{87}$$

The most commonly employed method for the generation of the nozzle contour is by the use of Foelschs Method as it is a purely analytical technique and the contouring proceeds in both direction around the inflection point i.e the section of maximum expansion angle. The detailed analysis of this and other relevant methods is presented in NACA-1651 [101].

The Figs. (29) & (30) below shows a slit nozzle developed for studies on a small scale COIL system by the authors,

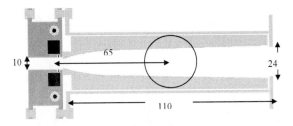

Figure 29. Integrated view of the Nozzle, Iodine injector and Cavity. All dimensions are in mm.

Figure 30. Photograph of the slit nozzle.

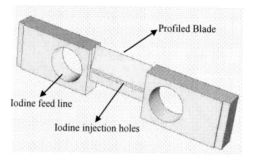

Figure 31. Typical blade of a grid nozzle.

Figure 32. Stacked nozzle blades forming a Grid.

Grid nozzles are actually a series of slit nozzles stacked together in order to provide the desired cavity conditions in high power COILs where the gain lengths are relatively large. A typical nozzle blade employed in a grid nozzle is shown in Fig (31) along with its stacked view in Fig (32).

Iodine injection also plays a critical role in the generation of the desired COIL active medium. The manner of iodine injection influences the iodine dissociation time, population inversion and thus the gain achieved. All these processes are primarily the function of the efficiency of mixing of the oxygen flow and the injected iodine. In case of subsonic COILs characterized by low pressure and low I_2 flow rates mixing is readily achieved. However, for new generation of COILs characterized by higher densities of O_2 $(^1\Delta_g)$ and supersonic flows in the cavity (requiring higher Iodine flow rates) [13] the problem of mixing becomes predominant. In particular, experiments [84] have shown that the complete O_2/I_2 mixing may not be achieved even at the far end of the laser section.

Qualitatively, mixing can be explained as follows. The iodine injected into the oxygen flow is concentrated in narrow jets. The jet cross section is initially (close to the injection point) much less than that of oxygen flow. Thus, the initial local number density of Iodine molecules is close to the order of magnitude of that of singlet oxygen, in spite of much lower Iodine flow rates. Thus, relatively higher initial concentrations of Iodine cause to competitive processes to occur; First is the acceleration of Iodine dissociation, due to an increase in number density of excited iodine atoms I^*, serving as the energy chain carrier for the dissociation reaction [102], the second is the fast quenching of the excited O_2 $(^1\Delta_g)$, I^*, I_2^* (vibrationally excited iodine molecules) by iodine molecules, and hence retardation of dissociation. The competition between these two processes governs the dependencies of the characteristic dissociation length and the maximum gain. Typically, for lower Iodine flows quenching of excited species is lower and hence the former process dominates and higher dissociation and gain are observed. For higher iodine flows the latter process dominates decreasing gain and destroying population inversion. Thus, dependencies of the dissociation length and the gain turn out to be non-monotonous, the minimum dissociation time and the maximum gain being achieved at some optimal value of Iodine flow rate.

Poor mixing affects the lasing power of the COIL leading to low lasing efficiency as with slow diffusion, the iodine is concentrated in narrow jets even in the cavity allowing most of the singlet oxygen flow passing through the cavity unreacted and also the higher concentration of the excited iodine atoms enhances quenching of O_2 $(^1\Delta_g)$ and pooling of I^*.

In general, the most common and convenient method of iodine injection is the cross flow injection into the main flow. The regimes of injection of iodine into the main flow could be subsonic, transonic and supersonic. The conventional method is to employ subsonic injection in Supersonic COILs. Here the general mechanics of jet injection is shown in Fig. (33)[103].

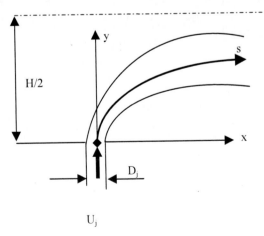

Figure 33. Configuration of cross-stream Iodine injection into the main flow.

The normalized jet penetration trajectories. y/D_j are correlated with the momentum ratio of the injectant and primary flows by,

$$P/d = C \left[\rho_s V_s^2/\rho_p V_p^2\right]^{0.5} \left[x/D_j\right]^{0.33} = Cq^{0.5} (x/d)^{0.33} \tag{88)'}$$

where, 'P' is the jet penetration 'C' is a constant which is equal to 0.56, 0.73, 0o.31 for the maximum concentration trajectory, and upper and lower and lower boundaries at 50 % of the maximum concentration trajectories, respectively. It can be easily seen that after a traverse of several jet diameters, the angle between the directions of the jet and the mainstream flow is close to zero i.e. the jet becomes parallel to the main flow. The penetration of the jet into the main flow is of the order of $q^{0.5}$ D_j. As the jet moves downstream the mixing process is visualized by the decrease in maximum concentration and from the spreading of the mass fraction profile. Downstream of the initial injecting location the mixing occurs via a shear – layer type process. The turbulent –like –shear layer mixing is simulated usually by nominal laminar diffusion coefficients, which are multiplied, using a Diffusion Coefficient Multiplier (DCM) [104], empirically adjusted to obtain agreement with experimental data. In general, the DCM can be expressed in terms of penetration height and injector hole spacing by,

$$DCM \propto (P_h/S)^2 \tag{89}$$

$$P_h = \left[\rho_s V_s^2/\rho_p V_p^2\right]^{0.5} \tag{90}$$

Where, 'P_h' denotes the penetration height, 'S' denotes the injector spacing, and ρV^2 denotes the momentum flow, and the subscripts s or p denote primary or secondary stream, respectively.

Typically, for RADICL device [103] employing a Cl_2 flow rate of 486 mmol/s with a dilution ratio (He as diluent here) of 3.265 with a nozzle height of 6.35 mm and an area ratio of 2.338 the penetration heights was 3.2 mm. The baseline penetration height was determined to be 3.1mm. The DCM value at a particular penetration can be determined by using equation (89) for both actual penetration and baseline penetration for the same injector hole spacing.

Another critical aspect in injection is the distance from the throat. The distance of iodine injection has a distinct effect on the power output obtained. Typical variation of COIL power output with variation in the point of injection with respect to nozzle throat is shown below [14] in Fig. (34),

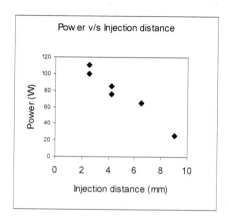

Figure 34. Variation of Power with variation in injection distance [14].

Also, the power obtained is also a function of iodine flow rates and it is found that the maximum laser power occurs for a certain optimum iodine flow rate. Both large iodine flow rates as well as small flows are detrimental to power output. In the former case most of the energy is lost in iodine dissociation thus reducing the available energy for population inversion while in the latter case the lower cavity concentration of the lasing species limits the power output.

3.3.1. Computational Analysis of Chemical Laser Flow Fields

The flow dynamics in a supersonic conventional COIL laser is known to be extremely complex [105] and a detailed analytical and theoretical modeling is highly challenging. Within the flow field of COIL there is multiphase flow, molecular diffusion, chemical non-equilibrium, thermal non-equilibrium and stimulated photon emission. In order to model the complete COIL lasing interaction, a three-dimensional formulation of the fluid dynamics, species continuity, and radiation transport equations is necessary. Even though various studies for COIL flow field have been carried out which have shown reasonable comparisons with measured gain, power and dissociation data, however, they have provided incomplete insight into quiet a few fluid dynamic aspects in the laser flow fields. This in part is due to two key reasons. First, the computational resources required to obtain detailed resolution of COIL flow fields are formidable. Second, the experimental data used to validate existing CFD/analytical models are incomplete for verifying important fluid dynamic assumptions, such as laminar or turbulent flow, and the unsteadiness within the COIL.

Miller et al. [105] have simulated a cross flow injection mechanism for better understanding of fluid dynamic phenomenon within geometries associated with COIL flow fields. They have employed a parallel, implicit, unstructured Navier-Stokes code Cobalt$_{60}$ for laminar and turbulent computations unsteady helium flows within the Research assessment device for improvement of chemical laser (RADICL). The studies reveal that most of the turbulence is observed in the region of the injector and a dominant unsteadiness, in the frequency range of 200 KHz, downstream of the nozzle throat is observed.

Carroll [106] has modeled a chemical laser flow field employing genetic algorithm technique. The technique has been employed to determine a set of unknown parameters that best matched the Blaze II chemical laser model predictions with experimental data. Significantly, mixing calculations carried out using the model indicates that higher iodine flow rates are necessary to maintain a nominal performance when high Helium dilutions are employed. The results were found to be in agreement with the experimental data collected on the RADICL.

Eppard et al. [107] have carried out one of the major efforts in the field of computational modeling of chemical laser flows and their optimization. They have developed a highly specialized flow sensitivity solver SENSE capable of handling reactive flows by incorporation of COIL chemistry model along with facilities for multi-component diffusion model with pressure diffusion terms, thermodynamic curve fits and modules to compute gain and gain sensitivity.

The sensitivity analysis requires a coupled fluid dynamic/ laser – optics resonator (LOR) system utilizing a geometric ray-tracing algorithm. The modules of SENSE and LOR are coupled to conduct a sensitivity analysis of a two-dimensional COIL geometry for cold and hot runs of the COIL.

$$\Re(Q, \beta) = 0 \tag{91}$$

The basic model utilized by the CFD formulation is that of residual map,

$$\frac{\partial \Re}{\partial Q} \frac{\partial Q}{\partial \beta} + \frac{\partial \Re}{\partial \beta} = 0 \tag{92}$$

Where, Q is the dependent variable of momentum, density, energy and β is the design variable. Thus, for a given design variable, the dependent variable is solved for by employing the eqn (91). Thus, a map is derived such that the flow solution is associated with a specific design parameter. Subsequently, for the sensitivity analysis the equation 92 is differentiated with respect to the independent variable β.

Finally, the equation 92 can be solved numerically with 'R' modeling the Reynolds averaged Navier- Stokes equation for turbulent flows employing finite rate chemistry. The sensitivity analysis is then coupled with the Laser-optics –resonator model. Generally, an implicitly-coupled system and generic design variable is incorporated as follows,

$$\Re_f(u_f, u_p, \beta) = 0 \tag{93}$$

$$\Re_p(u_f, u_p, \beta) = 0 \tag{94}$$

Where, u_p is the variable corresponding to power state and u_f represents the fluid state. In the ray trace optics model the relevant intensity fields are the initial field, the return field and the two way field (representing the coupling with the fluid dynamics of the system).

The initial and return fields are used as internal fields for the determination of the return intensity and then the two-way field for taking into account the effect of fluid dynamic conditions on the basis of current gain and initial intensity. The latter is determined using appropriate under-relaxation parameters. At convergence the initial intensity field and the return intensity field are the same.

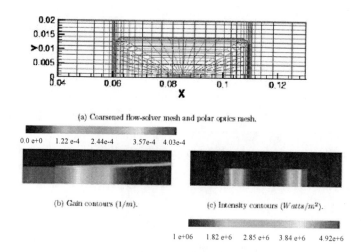

(a) Coarsened flow-solver mesh and polar optics mesh.

0.0 e+0   1.22 e-4   2.44e-4   3.57e-4 4.03e-4

(b) Gain contours (1/m).   (c) Intensity contours ($Watts/m^2$).

1 e+06   1.82 e+6   2.85 e+6   3.84 e+6   4.92e+6

Figure 35. Results of code used for simulating COIL flows [118].

A typical problem to validate the package was to test the sensitivity package on a grid topology of 151x51 for the flow solver embedded on to a 100x51 polar optics mesh. The inflow boundary was split into two lower and upper regions with I^* concentration (high gain region) and an upper stream with low O_2 ($^1\Delta$). The results obtained are shown in Fig. (35). The estimated power is 13.1kW.

3.4. Supersonic Diffuser

The basic challenge for the design of a pressure recovery system for a supersonic laser arises from the fact that the cavity pressures, in conventional systems, are of the order of 3-4torr, which are quite low. The purpose of the pressure recovery systems is to elevate the pressure from its value in the cavity to one that allows the laser gas to be exhausted to the atmosphere.

The specifics of the design of a pressure recovery system would be discussed in later sessions however at this stage it would be of interest to examine the kind of pressure recovery achievable via a diffuser. A diffuser basically works on the principle of increasing the pressure of the medium at the expense of the kinetic energy of the flow medium. The basic diffuser performance is governed by the equation [99],

$$P_x\left(1+\gamma M_x^2\right)=P_x\left(1+\gamma M_x^2\right) \tag{95}$$

Where, P_X is the cavity static pressure and the P_Y is the highest-pressure rise achievable by a diffuser if the flow is decelerated essentially to zero velocity ($M_Y = 0$). Thus, for typical COIL device conditions ($\gamma = 1.4$, $M_1 = 2$), the achievable pressure rise factor is ~ 6.6 thus the diffuser outlet pressure turns out be of the order of 20torr. Thus, it is evident that the passive diffuser pressure recovery cannot exceed the initial stagnation pressure ($P = P_o$) of the flow medium [99],

$$P_o = P\left[1+\frac{\gamma-1}{2}M^2\right]^{\frac{\gamma}{\gamma-1}} \tag{96}$$

In fact, this value is not attainable basically because of start / non-start phenomenon and unsteady flow phenomenon due to the occurrence of shock waves. Furthermore, the non-isentropic nature of the recovery process reduces the recovery even further. Although, the actual pressure recovery process in a diffuser via the occurrence of complex oblique shock wave phenomenon with the wall boundary, the entire process at best can be considered equivalent to a single normal shock, thus the maximum possible pressure recoverable is given by the following relation [99],

$$\frac{P_{oa}}{P_{ob}} = \frac{\left[\frac{\gamma+1}{2}M_a^2\right]^{\frac{\gamma}{\gamma-1}}}{\left[1+\frac{\gamma-1}{2}M_a^2\right]^{\frac{\gamma}{\gamma-1}}}\left[\frac{2\gamma}{\gamma+1}M_a^2-\frac{\gamma-1}{\gamma+1}\right]^{\frac{-1}{\gamma-1}} \tag{97}$$

Here P_{0b} is the pressure recovery possible through a supersonic diffuser but normally diffuser efficiency of around 0.8 [108] only is practically achievable. This decrease in efficiency occurs primarily due to flow distortion. From the above discussion it is evident that the diffuser alone is incapable of providing atmospheric pressure recovery.

Thus, the pressure recovery problem in COIL is complicated and can be treated by employing an active pressure recovery system such as an ejector in conjunction with passive recovery components. It is essential that the passive recovery component essentially a supersonic diffuser recover the maximum pressure possible thus minimizing the active system-pumping requirement. Thus, it is obvious that the diffuser design is as critical for COIL pressure recovery as is the ejector design.

A diffuser for conventional i.e. non-reactive flow with high Mach number consists of converging section upstream of the constant area section in order to decelerate the flow to subsonic regime. A divergent section follows the constant area section for further decelerating the high mach subsonic flow to nearly zero velocities.

In case of a chemical laser with low supersonic Mach number flow ($M \approx 2$) the constant area diffuser [109] are employed without incorporating the converging section since it enhances recovery beyond normal shock primarily in case of high supersonic flows only. Moreover, the flow being reactive has a self-tendency for convergence because of the decrease in flow area due to boundary layer growth occurring due to heat dumping.

Diffuser performance and size are important for efficient operation of COIL since it can adversely affect the cavity pressure and thus the laser output power. Moreover, size of a diffuser is another criterion since large diffusers pose problems in case of vehicular systems thereby compact systems are desirable along with maintaining optimal operation conditions.

Walter et al. [108] have given the empirical relations for the total diffuser length for high aspect ratio diffusers typical of these lasers taking in to account the unsteady flow phenomenon is given with relation to the characteristic dimension (A),

$$l = h^{0.3} b^{0.7} \qquad (98)$$

Thus, minimum diffuser length (L) for sufficient pressure recovery in COIL system is given as,

$$L \approx 7l \qquad (99)$$

An experimental study conducted by Walter et al. [108] has also outlined the basic problems regarding COIL diffuser design viz,

- Potential of cavity sensible heat dumping comparable to the laser power
- The actual heat dumping is strong function of the iodine flow rates, water vapor (liquid) flowed into the cavity form the SOG. These parameters tend to vary form run-to-run making diffuser design extremely difficult.
- Further, The accelerated heat release in the diffuser channels occurring due to coalitional deactivation caused by high water vapor density has the potential of thermal choking of the diffuser and causing excessively high cavity pressures.

The generic methodologies that have been suggested for overcoming these problems are;

- Extra expansion can provide pressure recovery with high heat release diffusers.
- Typically, it is expected that transverse vaned diffusers with an aspect ratio of 2.5 would perform better.
- Further, it is critical to trip the boundary layer in order to generate turbulent mixing but it should not have any notable effects on the laser conditions

Cold flow tests have only limited values for evaluating diffuser operation under laser flow regimes of COIL

It is quiet important to note that even a marginal underestimation of the cavity heat release or water vapor fraction could lead to impairing of the laser itself. Therefore, one is better off sometimes in not employing a diffuser in laser systems operating in low supersonic regimes (conventional COILs). Generally, in such systems a diverging duct is used after the cavity exit which perform distinctly better than an ill-designed diffusers both in terms of recovery and laser operation [108].

3.5. Trap and Vacuum Pumps

The trap at the downstream of the cavity is generally a liquid nitrogen cooled reservoir serving the purpose of trapping traces of chlorine and iodine being carried in the laser medium. These being harmful chemicals would not only adversely affect the material of subsequent sub systems but are also dangerous to be exhausted directly to the atmosphere. Moreover, it provides an additional volume in vacuum following the supersonic diffuser, which would aid diffuser start up phenomenon, further it converts the rectangular active medium output from the cavity to circular output which is best suited for ejector design.

The basic principle for designing the heat exchanger is that

$$\tau_r > \tau_d \tag{100}$$

i.e the residence time of the gas in the exchanger is greater than the specific diffusion time for the particle to be trapped. The individual values of τ_r & τ_d can be calculated from the following relations,

$$\tau_r = Q/V \tag{101}$$

Where, Q is the volume flow rate of the medium and V is the total trap volume available for gas flow.

$$\tau_d = (1/D)(V^2/S^2) \tag{102}$$

Where, D is the coefficient of diffusion for chlorine, S is the contact surface area. Hence, the condition for heat exchanger surface area is,

$$S > \sqrt{(V.Q/ D)} \qquad (103)$$

For chlorine $D = 0.001 \ m^2/ s$ at typical pressure of 10 torr corresponding to the recovery by the supersonic diffuser.

For a typical laser gas flow of 6.8 mole/s, consisting of all the gases including chlorine, buffer gas etc., at 10 torr (corresponding to diffuser recovery) and temperature of 80 K (T_{trap}) with a LNT gas flow volume of $0.2 \ m^3$, the volume flow rate can be estimated as,

$$Q = (P_{amb} \ T_{trap} \ G * 0.022.4)/ (P_{trap} \ T_{amb}) = 2.06 \ m^3/ sec \qquad (104)$$

Thus, substituting the above values for determining the surface area in eqn (103), the required surface area is found $20.3 \ m^2$. Usually a finned structure is used to realize such large areas.

COIL being a low-pressure high throughput system the total pumping requirement for the laser operation in the supersonic regime is quiet large. The total estimation of the pumping requirement per unit time can be estimated by utilization of the Boyle's law,

$$P_1 V_1 = P_2 V_2 \qquad (105)$$

Where, '1' represents the values of the laser medium at ambient conditions and '2' represents the corresponding values for the cavity conditions.

However, for the dump calculation the size of the dump required can be estimated by taking into consideration the total run length and in this case the 'V' terms would represent the corresponding total volume of gas flown for the specified time duration.

Thus, for a typical laser of 1kW power employing approximately 6gm/s of laser flows the pumping requirement of vacuum pumps would be $4500 \ m^3/hr$ and a corresponding dump requirement of $10 \ m^3$ for a 10 sec run period. Thus typical specific power output i.e. power per unit pumping capacity for conventional SCOILs is 1 J/litre.

Even though the roots blower can discharge directly to the atmosphere without any backing pump, it is not advisable in practice. In fact, a backing rotary pump makes it possible to operate the Roots pump at lower inlet pressure. Further, power required to drive the Roots pump is also reduced with backing pump. The main advantage of Roots pumps is their ability to handle large gas loads in which neither rotary nor diffusion pumps are fully efficient.

However, as has already been mentioned the specific power output of COIL is around 1 J/ litre leading large pumping requirements. Endo et al. [110] have produced a high-pressure subsonic COIL operation by not expanding the SOG flow thereby achieving high cavity pressure of ~ 6torr. Therefore, the pressure at the inlet of the vacuum pump is also ~ 6torr. Employing advanced turbo blower rather than conventional Roots blower further brought down the vacuum system size. The specific power output was improved by nearly 5 times using this method of COIL generation.

4.0. COIL: RECENT TRENDS

The COIL research has passed through many stages and the system development has now reached to such a maturity level where advanced countries are planning deployable systems based on this laser. Several dramatic changes has occurred in COIL research during these 27 years, and still require a lot to allow them to achieve their full potential as an utmost powerful laser source in the near infrared region. Much of the COIL researches so far have focused on efficient chemical singlet oxygen generators especially based on the liquid- gas SOG technology and have been successful to a large extent. However, this technique of generating the O_2 $(^1\Delta_g)$ has significant operational drawbacks, which includes the complexity in operation, corrosiveness and limited storage life time of the reagents and need for complex reactant feed systems: chemical pumps, tanks, chilling systems, water traps etc.. Introduction of all gas phase iodine lasers (AGIL) with the use of NCl $(^1\Delta)$ as the energy donor, which can be produced without a liquid phase generator and electrically assisted COIL are the major breakthroughs in this direction for both ease of operation as well as compactness. Further, the improvement of chemical efficiency can obviously reduce the size of the overall system corresponding to a particular power level. Recent trends can be classified into three categories viz. *efficiency and compactness oriented, new SOG approach and deployable systems*. The progress in these areas is briefly given in the following paragraphs.

4.1. COIL Efficiency and Compactness Issues

Investigation of technologies for the improvement of chemical efficiency of COIL and hence making the system compact is based on the heuristic equation given by Hon [22] as,

$$\eta_{chem} = U_{Cl}(Y_{\Delta,genexit} - Y_{\Delta,th} - Y_{diss} - Y_{loss})\eta_{mix}\eta_{extm}\eta_{extr} \qquad (106)$$

where, the notations have usual meaning as explained earlier. Here the chlorine utilization (U_{cl}) and the singlet oxygen yield at the generator exit ($Y_{\Delta,genexit}$) depend on the SOG performance and has almost reached a saturation with typical values of about 90-95% and 60-65% in most advanced SOGs. Also in typical COIL design, the singlet oxygen transport loss can at the most be reduced to about 2%. Moreover, much improvement cannot be made on the medium extraction efficiency (η_{extm}) and the resonator extraction efficiencies (η_{extr}). However, the techniques being pursued for efficiency improvement are relying upon the following factors:

1. Reduction of threshold singlet oxygen yield requirement by reducing the cavity medium temperature employing advanced nozzles of high Mach number
2. Improvement of mixing efficiency between the pumping and lasing medium flows using various iodine injection schemes
3. Introducing completely pre-dissociated iodine atoms instead of iodine molecules.
4. Improved singlet oxygen yield at the generator exit by using special fuels like BDP (Basic Deuterium Peroxide) instead of BHP

4.1.1. Different Iodine Injection Schemes

As has already been emphasized that mixing efficiency defined as the fraction of O_2 ($^1\Delta$) mixed with iodine is one of the critical parameters in COIL. In order to achieve high lasing power and improve efficiencies this parameter should be high. Another critical related factor is to reduce the O_2 ($^1\Delta$) losses during dissociation.

The classical subsonic schemes are often not optimal, in particular for high molecular weight buffer gas such as nitrogen. Replacement of helium buffer with nitrogen increases the molecular weight of the lasing medium nearly 3 times thus requiring that molecular iodine injection point to be moved closer to the throat. Since the local iodine concentrations in the subsonic part may be nearly of the order of O_2 ($^1\Delta$) concentrations [111] (which are distinctly much higher than the mean value after mixing), the maximum O_2 ($^1\Delta$) losses occur in the high-pressure subsonic region of the nozzle. Although, the scheme ensures high rate of Iodine dissociation which is desirable but it also accelerates quenching of I^* due to high molecular iodine concentrations, leading to medium heating and reduction in potentially extractable power. Thus, the contemporary researches in COIL involve exploration of various injection mechanisms and also possible generation of high-pressure gain medium in spite of the limitations on the generator side in terms of operating pressure.

Barmashenko et al. [25] have studied the effect of cross-stream injection secondary flows with Iodine in transonic and supersonic regimes of main flow without primary dilutions. The analytical model for determining the fraction of iodine dissociation (F) and the number of O_2 ($^1\Delta$) molecules utilized for dissociation reveals that the for typical F = 0.55 and with lesser time available for dissociation between the point of injection and resonator optics and the corresponding slow mixing rates of primary and secondary gases observed at higher Mach number injections the observed maximum gain in transonic injection schemes is found to be nearly 0.34 % cm^{-1} with mixing efficiency of ~ 0.5.

The efficiency of mixing in general for transonic and supersonic flows due to poor diffusion rates at higher Mach number conditions is expected to result in higher characteristic dissociation time, and the problem is further magnified by the delayed iodine injection thus reducing the available residence time for mixing and dissociation. Therefore, the only solution is to reduce the length scales of mixing and having a more uniform iodine injection. Therefore, one needs to be using more number of injection orifices for the secondary flow injection, which is actually limited by the geometrical considerations. Also, these nozzles can also be utilized as ejector nozzles by utilizing high-pressure secondary flows with or without primary dilutions. The experiments conducted with more uniform injection of iodine and also injections at various angles in order to evaluate the choking effects on the main flow have yielded mixing efficiencies of ~ 0.8 with a gain of nearly 0.6% cm^{-1}. The configurations of various supersonic injection mechanisms that have been used are shown below in Fig. (36),

The use of dense orifice network for iodine injection in supersonic regimes, though achievable, is not a complete solution to the mixing problems associated with these injection mechanisms. Another advance mixing concept is the use of parallel mixing at high supersonic velocities of pure oxygen and iodine + nitrogen. The concept also targets high stagnation pressure gain medium generation by utilizing high-pressure secondary flow. The large secondary flow is injected at a hypersonic Mach number of 5 and the singlet oxygen is injected under sonic conditions in order to achieve the required low-pressure conditions optimum for lasing [112]. The primary advantages that accrue from this kind of mechanism are jet mixing at supersonic velocities and low pressure considerably reduce the cavity temperatures, which in turn decrease the singlet oxygen threshold increasing the extractable power. Further, the detrimental losses are also reduced, as the iodine molecules are no longer injected into the singlet oxygen medium directly which may lead to higher local concentrations. The efficiency of a COIL system employing this nozzle is expected to be very high. Efficiency of the order of 33 % has already been demonstrated by Rybalkin et al. [113].

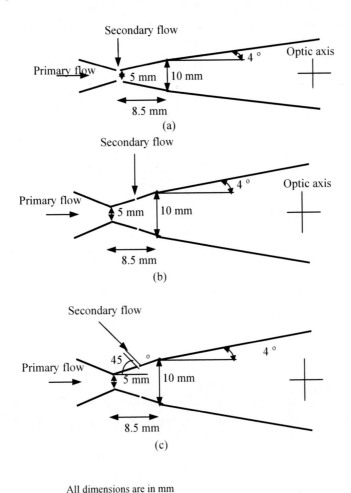

All dimensions are in mm

Figure 36. Slit nozzle (a) with transonic iodine injection, Slit nozzle (b) & (c) with supersonic injection of iodine.

However, one major concern with this approach expressed by various researchers is that parallel mixing of high velocity jets is distinctly poor due to high Mach number flows with low Reynolds number flow mixing stabilities are non - existent. Thus, complete iodine / oxygen mixing would require fairly large distances. This large distance would increase boundary layer thickness and significantly reduce the singlet oxygen yield. Again, one way of achieving complete mixing is by decreasing the distances between the centerlines of each nozzle pairs. This reduction reduces the diffusion distance, however, for complete mixing via pure shear would require extremely small sizes of the nozzle, which are difficult to develop due to fabrication limitations. Moreover, with small sizes the viscous effects tend to predominate accelerating the boundary layer growth and the loss of stagnation of pressure would correspondingly choke the flow thus reducing the pressure recovery potential.

Thus, the technique to overcome this problem is the use of vortex generators to reduce the length scales of mixing. The general method of producing vortices is to use trip jets i.e flow injections, but trip jets are inefficient at low pressures and are also configurationally complicated. Thereby, use of tab arrays arranged at an optimum angle of attack, typically 10 deg, have been suggested to generate an extensive network of stream wise vortices leading to efficient mixing. The tabs need to be optimized in terms of their lengths and the inter tab spacing. Smaller tabs in general tend to prove inefficient as they produce vortices much closer to the nozzle walls, which find it difficult to diffuse into high momentum nitrogen region. Also, smaller spacing tends to cancel out the vortex strengths and reduce flow entrainment. The typical *tab arrangement* (Fig. (37)) and the nozzle blade configuration of such nozzles is shown below,

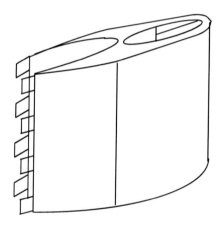

Figure 37. Advanced nozzle developed by Boeing [111,112].

The supersonic mixing is one of the current interests in the field of COIL for high efficiency operation due to reasons of reduction of O_2 ($^1\Delta$) losses cited earlier but also as conventional supersonic injection is not suitable of atomic iodine injection scheme. When atomic iodine is injected instead of the molecular one and mixed at the nozzle plenum, quenching reaction occurs at a very high rate and large amount of energy is lost prior to supersonic expansion. Thus, injection in the low-pressure regime is essential for successful atomic iodine injection [114]. We have already emphasized the need to reduce the length scale of mixing in order achieve sufficient gain at the optical axis thereby requiring some kind of vortex generation methods. Endo et al. [5] have proposed a *winglet nozzle concept* in order to improve mixing and thus achieving high efficiency laser operation.

The basic principle of these winglet nozzles is the use of the concept general termed as the lobed mixers. The typical geometry of a lobed mixer is shown in the Fig. 38.

In a lobed mixer [115], the generation of stream wise vortices is associated with the variation in the aerodynamic loading along with the span of the mixer, analogous to the situation along a finite wing. At the trailing edge, a continuous distribution of stream wise vortices is discharged into the flow, evolving downstream into an array of discrete counter-rotating vortices, as shown above. The vortices grow through turbulent diffusion and circulation eventually decays as the counter rotating vortices diffuse into one another. The stream wise vortices downstream of such forced mixing devices are typically larger, in both magnitude and scale, than those in naturally developing free shear layers and boundary layers (typical of mixing in flows without vortex generation).

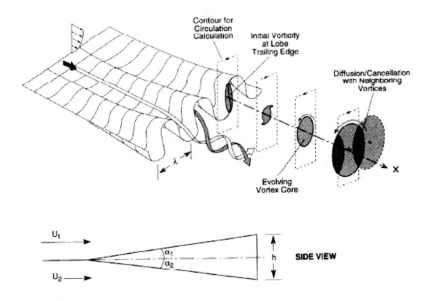

Figure 38. Mechanism of vortex generation in lobed/ winglet mixers.

A good estimate for initial stream wise circulation due to the lobes is given by assuming that the fluid, which exits the lobes, does so at the lobe angles. For, the geometry shown, this yields a shed circulation of magnitude 'Γ'

$$\Gamma \propto U_1 h \tan \alpha_1 + U_2 h \tan \alpha_2 \qquad (107)$$

Where, U_1 and U_2 are the free stream velocities on either side of the lobe, h is the lobe height and α_1 and α_2 are the lobe penetration angles. A constant 'k' whose value depends on the lobe geometry is necessary in the scaling to provide a quantitative estimate of circulation. For, the mixer with vertical sidewalls, the value of 'k' is unity. Although, these mixers also produce transverse vortices along with the stream wise ones, but the length scales of the former are set by the shear layer thickness at the trailing edge whereas the latter are governed by the half-wavelength of the lobe geometry, $\lambda/2$. Typically, the boundary layer thickness at the trailing edges is much smaller therefore the scale of transverse vertical motions is much smaller as compared to that of stream wise vortices for a distance downstream of approximately three to ten wavelengths.

Thus, a geometry similar to the above shown has also been used in the form of X-wings [116] (as shown in Fig. (39)) for the generation of stream wise vortices.

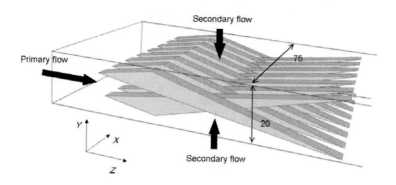

Figure 39. Schematic drawing of the X-wing type supersonic nozzle for COIL.

The winglets are placed at an angle of attack of 20° and the duct height is 20 mm and a width of 75 mm. The use of this form of mixing with medium Mach number of 1.9 has yielded a power of 599 W with an efficiency of 32.9 %[5]. This is the highest reported efficiency for Supersonic COILs till date.

4.1.2. Atomic Iodine Generation Schemes /Hybrid Electri-COIL

In conventional COIL, a fraction of the energy carriers is spent for the dissociation of the iodine molecules in forming the lasing iodine atoms. Therefore, the availability of these energy donors for pumping the iodine atoms will reduce and hence decrease the efficiency of the system. Alternative is the direct supply of the dissociated iodine atoms into the singlet oxygen flow. It has been estimated that a production of atomic iodine without participation of singlet oxygen could increase a laser power by 25%[117].

There are three approaches to generating atomic iodine: (i) Direct dissociation of I_2 by electrical means (i.e., microwave/electric discharge)[118], (ii) bond splitting of iodine donor molecules like CF_3I or CH_3I by means of photolytic/RF discharge [119], and (iii) replacement of iodine atoms from reactions involving an iodine donor molecule and atomic halogens like F or Cl [120].

4.1.2.1. Iodine Dissociation by Electrical Means

Wakazono et al. [121] used an RF discharge (50.3MHz) for the pre-dissociated supply of iodine. In their set up, the iodine molecules along with argon is passed through an RF discharge cell placed between the iodine injector and bubbler SOG [121]. In their experiments they found that the chemical efficiency improved from 4% (in their normal COIL operation) to 12% in RF discharge case and justified that in case of normal operation the molecular iodine was not perfectly dissociated by singlet oxygen molecules. Since their efficiencies are very low compared to the other existing COIL systems, it looks that the 3 times improvement in their efficiency is not only because of the iodine dissociation but also because of the improved conditions like operation at higher iodine concentrations (as compared to normal operation) etc. Endo et al. [117], demonstrated the transonic and supersonic injection of pre-dissociated iodine by means of microwave discharge in a COIL set up, the scheme is shown in Fig (40), with the two separate nozzle configurations: slit and ramp respectively [114]. The degree of iodine dissociation in these experiments is 20-40% and depends on the iodine concentration. The authors have reported the maximum improvement of 9 % in chemical efficiency.

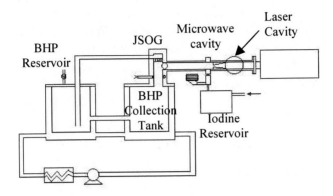

Figure 40. The schematic of a hybrid Electri- COIL.

4.1.2.2. Bond Splitting of CF₃I or CH₃I by Electric/Photolytic Discharge

The above mentioned direct methods for dissociation of I_2 molecules into atoms require gaseous I_2 which has its own associated thermal management issues especially while going for high power systems. An alternative method to produce iodine atoms by decomposing methyl iodide employing a dc glow discharge in a vortex gas flow has been suggested by Mikheyev et al. [119]. With this technique they could demonstrate an iodine concentration of 3.6×10^{15} cm^{-3} at a pressure of 15torr with argon. This concept of premixing an iodine donor molecule (methyl iodide, for example) with the primary oxygen flow and liberating the iodine atoms using RF discharge or UV photolysis resulting in a uniform mixture of iodine and singlet delta oxygen is extremely attractive.

4.1.2.3 'I₂' Atoms from Reaction between I₂ Donors & Atomic Halogens

One of the alternative approaches to produce atomic iodine generation is to generate them through chemical reactions as given below [122]:

$$F (Cl) + HI \rightarrow HF^* (HCl^*) + I (^2P_{3/2}) \quad k_f = 6.3x10\text{-}12 \; cm3/s \qquad (108)$$
$$k_{Cl} = 1.64x10\text{-}10 \; cm3/s$$

Under optimum conditions, Spalek et al. could demonstrate the production of small-scale atomic iodine with yield greater than 70 %[123]. Recently, Hewett et al. [124] have used the proven combustor technology (developed for HF/DF lasers) to produce large quantity of Fluorine atoms for the above said reaction. These generated fluorine atoms are introduced along with HI molecules to produce atomic iodine. The conversion efficiency of about 75% has been reported using this technique.

4.1.3. Utilizing Basic Deuterium Peroxide (BDP)

The major limitation of COIL chemical efficiency as from equation (106) is decided by the singlet oxygen yield one is able to get at the singlet oxygen generator exit. The typical value of this singlet oxygen yield is about 65% and the reduction is mainly because of its quenching in BHP liquid inside the generator. BHP is an electrolyte containing many ions and molecules and especially H_2O, H_2O_2, HO_2^- and Cl^- are the strong quenchers for singlet oxygen. Since the concentration of water molecules in BHP is about an order of magnitude larger than the other quenchers, the effect of water is considered to be the most significant one in case of BHP based system. The experiments have shown that the quenching of excited iodine atom by H_2O and H_2O_2 molecules is nearly 50 and 3 times higher compared to that of D_2O [125] and D_2O_2 molecules [126] respectively. Vetrovec et al, have carried out a detailed theoretical analysis and shown that the replacement of BHP with Basic Deuterium Peroxide (BDP) can improve the detachment yield from 65% to more than 85% and the COIL chemical efficiency from 30% to 45%. [127].

4.2. Advanced Generators for Excitation of Iodine

The present day COIL systems are based on the liquid-gas phase reaction for the production of singlet oxygen. The major drawbacks in these systems are the size, operation problems, handling of liquid chemicals and the start up time. In order to counter these disadvantages, radically new singlet oxygen generators have been tried for the development of these lasers. The major breakthrough obtained in this direction is the demonstration of the All Gas Iodine Lasers by Air Force Research Laboratory (AFRL), USA, where about few tens of watts of laser power has already been demonstrated with $NCl(^1\Delta)$ generated from chemical reactions involving only gas phase mediums as the pumping source instead of $O_2(^1\Delta)$ [61]. The second in this direction is the Electric COIL in which $O_2 (^1\Delta)$ is generated through electrical means. Brief details about these technologies and their current status are discussed in the following sections.

4.2.1. All Gas Iodine Lasers

Benard [40] has suggested that the singlet nitrenes such as $NF(a^1\Delta)$ and $NCl(a^1\Delta)$ can serve as a possible replacement for the singlet oxygen due to their similar electronic configuration. These molecules also have moderate lifetime, 2-5 sec and hence can be potential candidates as the energy onors for COIL applications. The NCl(a1Δ) has higher potential compared to the $NF(a^1\Delta)$ due to its fast reaction with iodine atoms in accordance with the relation[40]:

$$NCl(a^1\Delta) + I \rightarrow NCl (X^3\Sigma) + I^* \quad k = 1.5 \pm 0.7 \times 10^{-11} \, cm^3/ \, molecules/s \quad 109)$$

The NCl(a1Δ) molecules are produced from fluorine atoms and hydrogen azide (HN_3) by a three step process,

$$F + DCl \rightarrow DF + Cl \quad\quad\quad\quad\quad (110)$$

$$Cl + HN_3 \rightarrow HCl + N_3 \quad\quad\quad\quad\quad (111)$$

$$Cl + N_3 \rightarrow NCl(a^1\Delta) + N_2 \quad\quad\quad\quad\quad (112)$$

Recently about 15W of AGIL power has been demonstrated by Manke et al. employing $NCl(a^1\Delta)$ pumping with a possibility of power scaling up [61]. The typical AGIL scheme is shown in Fig. (41).

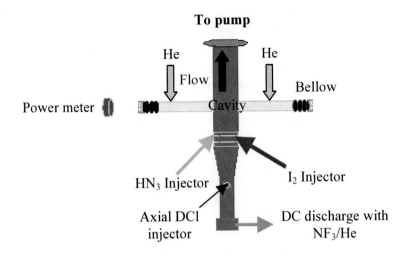

Figure 41. Typical Scheme for AGIL system.

4.2.2. Electric COIL /Discharge Oxy-Iodine Laser (DOIL)

An electric discharge passing through pure oxygen or oxygen rich gas mixture can result in the production of the singlet oxygen by the interaction of the molecules with electrons. Many workers [128-132] have attempted this technique, however, achieving positive gain has been the biggest challenge. Hill produced singlet oxygen with a yield of 15 % using a Controlled Avalanche Process. In this technique, a short high voltage pulse initially produces ionization while a comparatively low electric field supports the electric current between ionizing pulses [129]. Fujji and his team [131,132] from Japan have used RF discharge in a gas flow (both subsonic and supersonic) for the production of singlet oxygen molecules. They have reported singlet oxygen yield of 32% using this technique after proper optimization of gas mixture and electrode configuration. Even though the level of singlet oxygen concentration seems to be of practical interest the challenge to be met is to achieve singlet oxygen at higher pressure, which at the present moment is only few torr [133].

Napartovich et al. [133] have carried out a detailed theoretical analysis of the processes involved in the production of singlet oxygen through electric discharge technique. In a discharge, an energetic electron collusion with the species and their momentum transfer to the species results in generation of higher energy molecules or occurrence of dissociation and ionization depending on the electric field strength and cross-section for the process. The gas at the exit of the discharge generator is highly excited, dissociated and heated. In addition, the ozone is effectively formed which is a strong quencher of singlet oxygen molecules. Thus, achieving high singlet oxygen yield is a complex issue and requires optimization of electric field strength (E/N). The theoretical studies indicate that an optimum E/N value of about 1×10^{-16} Vcm^2 is required for obtaining high singlet oxygen yield and suppress the unwanted processes which add additional heat and contribute towards pooling losses.

Using mixtures of oxygen with some gases results in variation of the electron energy balance. Generally, one may expect a decrease of the electron energy fraction going into direct excitation of $O_2(^1\Delta_g)$ when molecular or atomic additives appear. But modeling and experimental results show an interesting positive result. These include stabilization of the plasma, increase in discharge efficiency, lowering the gas temperature etc [134]. Recently, Ionin [134] has reported that a yield of 25% is theoretically possible at high pressures using a gas mixture of O_2: Ar: CO:: 1:1:0.1 employing electron beam sustained discharge technique.

Further, employing electrical methods for generation of $O_2(^1\Delta_g)$ molecules there are added advantages as well. Firstly, there is generation of sufficient amount of $O_2(^1\Sigma)$, which is useful for dissociating iodine molecules. Secondly, the absence of water vapor improves the efficiency of the system significantly. Thus the chemical efficiency will be higher in case of Electri- COIL as compared to the existing classical COILs for the same singlet oxygen yield [53].

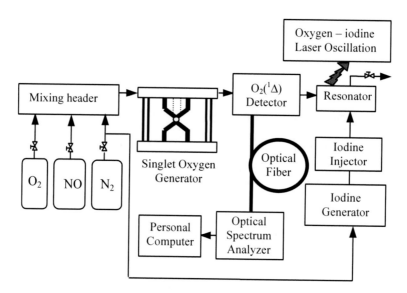

Figure 42. Typical Scheme for DOIL system.

The DOIL experiments have been conducted using rf, microwave and dc discharges. A typical configuration for a discharge oxygen iodine laser is shown in Fig. (42). Rawlins et al. [135] and Carroll et al. [18] have reported a positive small signal gain in subsonic flow reactor in their initial experiments. Even though these reported gains are smaller as compared to the present day classic COILs, but one should compare these results with those observed during the initial period of classic COIL operations.

Very recently, Carroll et al. [62] have demonstrated the first lasing action from an electric COIL using rf discharge technique. Maximum power of 220 mW was observed when about 450 W of rf power was fed in the frequency range of 1700 MHz.

4.2.3. Fullerene –Oxygen Iodine Laser (FOIL)

One of the recent advances in Oxygen –Iodine Laser research in which success would lead to an alternative manner of chemical production of singlet oxygen is the Fullerene based

Oxygen Iodine Laser (FOIL). The idea of FOIL initiated by Zaleskii [136] in 1983, but the first experiments could not successfully demonstrate lasing action.. The singlet oxygen molecules are generated through the chemical kinetics between the oxygen molecules and the fullerenes, which are optically excited through irradiation using a pulsed light or even simulated solar radiation [137]. It involves the following kinetics for the production of singlet oxygen. The fullerene in the ground state (^0F) absorb a photon and convert into their singlet state (^1F) with subsequent non radiative excitation transfer to the meta-stable triplet state (^3F) with an effective cross-section of, $\sim5\times10^{-18}-10^{-17}$ cm^2 as,

$$h\nu + {}^0F \rightarrow {}^1F \rightarrow {}^3F \tag{113}$$

The interaction between these triplet states of fullerene and oxygen molecules at ground level can result in singlet oxygen molecules production as

$$^3F + O_2 \rightarrow {}^0F + O_2(^1\Delta) \quad k = 3.3\times10^{-12} \text{ cm}^3 \text{ s}^{-1} \tag{114}$$

Many variants have been used for the production of singlet oxygen molecules based on the above principle. These include a multilayer solid-state membrane of fullerene irradiated with photon flux and a fullerene solution/ suspension boiling under irradiation of light [138-140]. Both these techniques have not resulted in sufficient quantity of singlet oxygen required for lasing action.

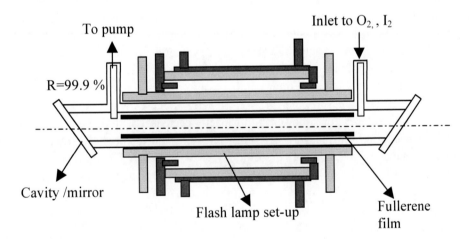

Figure 43. Schematic of FOIL[60].

Recently Belousova et al. [60] have succeeded in achieving laser action through this technique. The set up used by the authors is shown in fig (44), where the inner surface of the cell (quartz cylinder) has been coated with a thin film of fullerene (C60) and is optically pumped from outside using a flash lamp. With pump energy of 0.6 kJ, 50 cm long and 1 cm diameter cylinder cell having a fullerene coating density of 3×10^{17} particles/ cm^2, the authors were able to produce laser pulses of 25mJ energy.

4.3. Deployable Systems: Development of Pressure Recovery System (PRS)

COIL is a low pressure and high gas flow system. This in turn requires high capacity vacuum pumps, which makes the COIL system bulky. Serious efforts are being made to develop pressure recovery system that can provide the isolation between the low-pressure cavity and laser exit along with achieving direct atmospheric discharge of the laser medium. There are two approaches being followed, first one is based on sorption whereas the second one is based on ejector. The sorption-based system has been demonstrated on a small scale. Although, the scheme is workable and is capable of providing compact COIL systems but has limitations regarding field deploy ability as the systems work well at lower temperatures and also require substantial time for regeneration of the Zeolite bed thus increasing the lead time for laser availability. The alternative for making COIL as a compact system by avoiding the evacuation system is to use an ejector based pressure recovery system [151]. The US Air force is in the process of developing COIL laser of ~200 kW power on a Boeing Airbus [7] using ejector based pressure recovery system. These approaches are discussed in brief in the following paragraphs.

4.3.1. Zeolite Vacuum Absorption Bed

Cryosorption pumps of various designs have been used in the vacuum industry for many years. Sorption pumps function by the physical adsorption of gases at the surface of molecular sieves (Zeolite) or other sorption material (e.g., activated Al_2O_3 or charcoal) [142]. Such materials have an extraordinarily large specific surface area per unit of mass (~1000 m^2/gram). Correspondingly, the capability of gas adsorption is considerable, around 200 milligrams of N_2 per gram of synthetic zeolite (Linde 4A, at the temperature of liquid nitrogen (77 K) [143]. A variety of natural and synthetic zeolites are now commercially available [142]. Sorption capacity of zeolites (maximum amount of gas that can be stored) is highly dependent on zeolite temperature and pressure of gas above the sorption surface. In particular, at a constant pressure, the sorption capacity increases with decreasing temperature while at a constant temperature, the sorption capacity decreases with decreasing pressure. For example, at an equilibrium pressure of 10torr, changing the temperature from 293 to 77 K increases the capacity of zeolite (e.g., Linde 4A) to adsorb N_2 to more than 200 times. Furthermore, during the sorption process the sorption effect decreases with increased coverage of the sorption sites.

The pump consists of a gas chiller and a bed of zeolite. Gas exhausted from COIL laser cavity is first chilled to about 100 K and condensable vapors are removed. Cold and dry gas is then adsorbed onto a bed of zeolite cooled to a temperature of approximately 80 K. When the sorption capacity of the zeolite is exhausted, by allowing the zeolite and the gas chiller to warm up and liberate trapped gases can regenerate the pump. Gases liberated in the regeneration process are removed by a small mechanical vacuum pump. The cryosorption pump is a compact lightweight device, which can provide effective vacuum pumping for transportable COIL systems. Gas adsorption process is accompanied by a significant release of thermal energy known as the heat of adsorption (H_a). This heat of adsorption is deposited into the zeolite, resulting in increase of its temperature thus reducing its sorption capacity. To maintain sorption capacity, removal of the adsorption heat may be handled in two distinct ways. In the first approach, the zeolite bed can be operated in an inertial mode. Here, the amount of zeolite is increased to allow it to soak up the heat of adsorption with only a modest temperature rise and with some tolerable loss of sorption capacity. In the second approach, heat of sorption is removed from the zeolite in real time. Low thermal conductivity of zeolites makes such process a challenging task and also impedes fast regeneration of the cryosorption pump [144,145].

A COIL integrated with a cryosorption vacuum pump system is shown in Fig. (44) [146]. The system uses N_2 diluents and the pump adsorbs all the laser cavity gas flow. The pump system comprises a gas chiller/condenser, zeolite bed in a vacuum enclosure, refrigerant feeds, means to warm-up the gas chiller and zeolite bed, and auxiliary mechanical pumps. During laser operation, decelerated flow from the diffuser enters the cryosorption vacuum pump system through the isolation valve 1. The flow is directed into the gas chiller/condenser where it is chilled to approximately 100 K. This process condenses chlorine, iodine, and water vapor and traps them onto condenser surfaces. Cold and dry gas containing only nitrogen and oxygen flows from the gas chiller into the zeolite bed. The bed uses a suitable sorption material such as the synthetic zeolite Linde 4A cooled to a temperature of ~80 K to adsorb and trap the incoming mixture of nitrogen and oxygen gases.

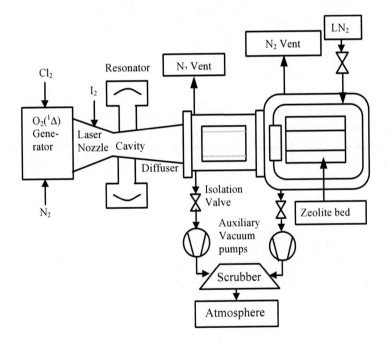

Figure 44. Schematic of Zeolite based pressure recovery system.

The gas chiller and the zeolite bed are maintained at their operational temperature by LN_2 cooling. The supply of LN_2 must be sufficient to cool down the gas chiller/condenser and zeolite bed, overcome thermal leaks, and reject the heat released by cooling and condensing and sorbing laser gases. A heat shield positioned between the Zeolite bed and the vacuum enclosure reduces radiative and convective heat load to the zeolite bed. In some cases, maintaining certain backpressure at the end of the diffuser could be desirable. For this purpose, an automatic throttle valve may be installed downstream of the diffuser. Typically for a 23.7 kW laser using nitrogen dilution, the requirement of zeolite and liquid nitrogen is of the order of 2698 kg and 1306 kg respectively for one hour of operation. However, this requirement reduces to less than one fifth, if one uses cryosorption pump that employs a closed loop that circulates helium diluents between COIL and zeolite bed [146].

4.3.2 Ejector Based Pressure Recovery

COIL systems for their compactness require an efficient pressure recovery system at the downstream of the diffuser. Ejector based pressure recovery system seem to have potentials of providing effective pressure recovery and are basically passive pumping devices. Ejector employs a motive jet with a large kinetic energy to induce a secondary stream to flow through a constant area or a converging mixing section. Ejectors typically fall into two main categories. The first of these has a large ratio of the suction area to the motive area thus producing a large entrained flow. Furthermore, it is not able to pump the suction flow against a large pressure rise. Thus, these types of ejectors can be termed as 'mass augmenters'. The second kind of ejectors are those in which the ratio of suction area to primary area is small thus leading to an appreciably low entrained flow but capable of pumping the suction stream from a low stagnation pressure to a sufficiently high backpressure. These types of ejectors are generally classified in the category of 'jet pumps' [147]. It is quite evident that in a COIL system one would need to employ the kind of ejectors classified in the latter category.

An ejector or a jet pump functions by expanding a high total pressure motive gas fluid through a nozzle or a set of nozzles into a low-pressure suction fluid and entraining it. In an ejector system the motive gas imparts its energy to the low energy suction fluid thus raising its pressure to provide the necessary pressure recovery. Thus, an ejector can be likened to a turbine –compressor system (Fig (45)) in which the motive gas functions as a turbine expanding from a high pressure to a low pressure performing work which is utilized to compress the suction fluid from a low inlet pressure to a higher exit pressure. The, process can ideally be considered to be an isentropic process [148].

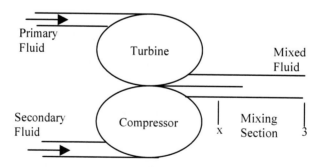

Figure 45. Ejector as a turbine-compressor system.

The number of stages an ejector needs to employ for exhausting the laser medium to the desired backpressure. Thus if a reasonable compression ratio of '10' is achieved in each stage the number of stages can be determined as,

$$P_{amb}/P_{diffexit} = (R)^n \qquad\qquad (115)$$

Where, P_{amb} is the ambient exhaust pressure, $P_{diffexit}$ is the exit diffuser pressure (~ 10torr). Thus, it is evident that for a conventional COIL system a two-stage ejector should suffice the requirement of pressure recovery.

Another, important aspect in ejector design is the choice of motive fluid and their supply pressure and temperature conditions. Generally, the entrainment ratio (suction to motive gas flow) for a given compression ratio increases with increase in supply temperature (proportional to square root of temperature ratio) and also with increases in specific heat ratio of the gas [149]. Another, important factor, is to consider the aspects of decreasing load for the high-pressure ejector stage. This can be accomplished by using steam as the motive fluid, which can be condensed at the end of the low – pressure stage using appropriate spray condensers. Steam also has advantages of in terms of its possible availability at higher temperatures. However, steam generation in a compact manner is another major task, as conventional steam generation methods do not provide the necessary compactness for transportability. Thus, decomposition of high concentration H_2O_2 (90% concentration) with silver catalyst has been proposed solution for compact steam generation [150]. The Airborne laser program [7] is also believed to employ a steam based ejector system with H_2O_2 decomposition used for generating steam. However, for continuous operation of COIL particularly suitable for industrial applications, safe handling of concentrated hydrogen peroxide is a prime issue. Thus, in design of COIL pressure recovery one needs to take into account the various aspects of motive gas type, generation and supply mechanisms and make a choice on the basis of the requirements and the deployability of the laser system.

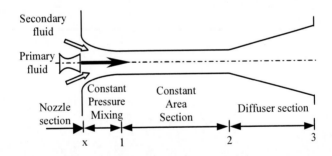

Figure 46. Conventional central ejector configuration.

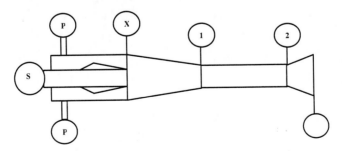

'S' Cavity or suction flow; 'P' Primary or Motive flow
'X' Mixing chamber conditions;'1' Inlet of constant area duct
'2' Exit of constant area duct; '3' Exit of subsonic diffuser

Figure 47. Peripheral ejector configuration for COIL.

Further, one of the most critical aspects in ejector design for a COIL system is the geometry of the ejector. Conventionally, ejectors employ central injection schemes (Fig. (46)), however, for COIL peripheral injection systems have been suggested as the most suitable ones. This is because the peripheral injection reduces the problems associated with the boundary layers [141] and thus helps in smooth laser operation. Thus, a general configuration of the peripheral ejector is shown below in Fig. (47).

The intricacies in design of an ejector can be better understood by looking at a design for a 500 W COIL system .

The COIL system under laser run conditions develops a cavity pressure of ~ 3torr with a cavity Mach number of ~ 1.5. The flow from the nozzle exit (75x 24 mm^2) expands into a duct with ϕ 75 mm. The important parameters estimated for two stages employing isentropic relations are listed below:

First Stage ejector

Entrainment Ratio	: 0.0025
Suction flow	: 3 gm/sec
Motive flow	: 120gm/sec
Motive supply pressure	: 2.0 bar
Mach number	: 4.85
Annular duct diameter	: 120 mm
Constant area duct diameter	: 104 mm

Second Stage ejector

Entrainment Ratio	: 0.03
Suction flow	: 123 gm/sec
Motive flow	: 4000 gm/sec
Motive supply pressure	: 38 bar
Mach number	: 4.87
Annular duct diameter	: 200 mm

One of the critical issues in supersonic ejector design is to achieve a near constant pressure mixing inside the mixing chamber, for which the convergence is determined from the Eq. (117)[151],

$$\tan \frac{\theta}{2} = \frac{\left(\sqrt{\frac{a_x}{a_2}} - 1 \right)}{\left(\frac{L_1 + L_2}{D} \right)} \tag{116}$$

The convergence angle employing the above relation turns out to be ~ 2.4degree and the lengths of converging and straight ducts are nearly 6D each. These correspond to the condition of achieving proper mixing and generation of transverse shock required for pressure recovery. Therefore, the first stage has a designed static recovered pressure of ~ 70 torr and an exit Mach number of ~ 0.4. The first stage ejector is connected with the second stage via a subsonic diffuser. The developed hardware of the first stage of the ejector module is shown in Fig. (48).

Figure 48. Developed first stage ejector hardware.

The computational studies on the module also provide considerable insight into the undergoing mechanism on these kinds of ejector systems. Figures (49) and (50) show the temperature and Mach number variation in the first stage nozzle using the specified boundary conditions for a considered mean surface roughness of (k) of 0.0025 for the evaluation of the friction factor corresponding to the existing local Reynolds number.

The turbulence intensity is taken as 10 % of the inlet kinetic energy and the length scale is taken as 7 % of the inlet characteristic dimension.

Computationally, in an ejector it is found that the gas at low Mach number from main flow inlet mixes with the supersonic flow from the nozzle over a region of rapid Mach number change where the mixing takes place, subsequently the resulting mixed stream is fully supersonic. The Mach number is found to decrease continuously in the converging duct and also in the constant area duct. The flow turns subsonic via a long oblique shock wave, which stands close to exit of the constant area duct. Fig. (51) shows the variation of static pressure along the axis of the first stage ejector and the location of a long oblique shock. The predicted pressure downstream of the shock in the first stage is ~ 63 torr.

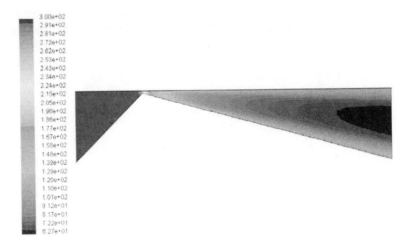

Figure 49. Temperature variation along first stage nozzle.

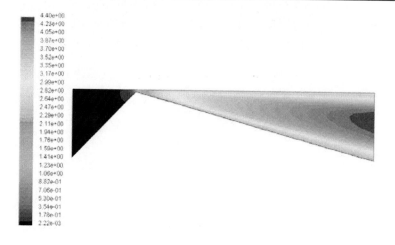

Figure 50. Mach number variation along first stage nozzle.

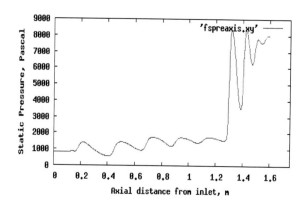

Figure 51. Static pressure variation along the axis of the ejector.

For a typical operation duration of 5 sec with the slit nozzle for an initial set backpressure of 60 torr in vacuum dump, the observed temporal variation of the cavity and Pitot pressure are shown on a logarithmic scale in Fig. (52). The observed cavity pressure and estimated Mach number are ~ 3 torr and ~ 1.5 respectively.

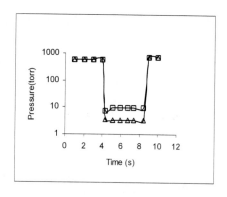

Figure 52. Temporal variation of Pitot and cavity pressure on a Logarithmic scale.

The designed motive gas flows in the first and second stages respectively along with the simulated laser gas flows were flown to test the complete ejector module. The ejector module was operated successfully for a start up pressure of 605 torr, with the dump pressure rising to ~ 715 torr by the end of the run without experiencing shut off (refer Fig. (53))

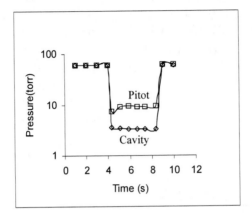

Figure 53. Variation of Cavity and Pitot pressure with time, log (p) v/s time. 'Δ' corresponds to cavity pressure, '□' corresponds to Pitot pressure.

4.3.3. Advanced Nozzles for Efficient Pressure Recovery

Nikolaev et al. [152] have suggested advanced nozzle known as ejector nozzle that not only provide conditions for excellent gain but also generate high stagnation pressure, which is conducive for better pressure recovery. The conditions achieved via such type of nozzles are extremely favorable for the design of single stage ejector based pressure recovery system for direct exhaust in COIL as well as in similar systems operating at low-pressure conditions. The nozzle consists of separate choked injection openings for singlet oxygen. COIL power of nearly 700 W has been demonstrated using this nozzle with an efficiency of 19.7%. The high Mach number conditions of ~ 2.5 under a dilution of 1: 11 have shown a maximum achievable gain of 0.7 % cm^{-1} [152].

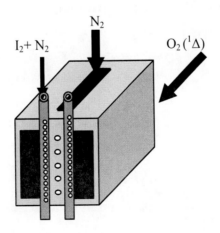

Figure 54. Fragment of nozzle bank.

A fragment of the nozzle suitable for 500 W class COIL is shown in Fig. (54).

It consists of 9 parallel slits of 10 mm height and 3mm width for the injection of oxygen flow from the JSOG carrying the pumping media. The high momentum flow of pure N_2 at high pressure is injected via 50 cylindrical nozzles of 1mm diameter spread over 10 rows. The N_2+I_2 flow is injected via holes drilled in 18 brass tubes interspersed between oxygen and nitrogen nozzles, (refer Fig. (55) & Fig. (56)).

Thus, typically for a 500W class COIL system, the design calculations were carried out for the constituent stagnation pressure conditions of 25 torr, 1450 torr and 120 torr for singlet oxygen, primary nitrogen and secondary nitrogen corresponding to flow rates of 371 mmol/s, 22 mmol/s and 27.5 mmol/s respectively. Each of the gas constituents expands to a common pressure of 10torr. The typical Mach numbers for singlet oxygen, primary diluents and secondary diluents were 1.3, 3.15 and 1.1 respectively. The overall computed values are shown below,

Mixing chamber area	7.5 cm^2
Mixed stream pressure (supersonic regime)	9.5 torr
Mixed stream Mach number	2.6
Mixed stream pressure (subsonic regime)	88 torr
Mixed stream Mach number	0.5

Figure 55. Photograph showing singlet oxygen slits & Primary Nitrogen injection holes.

Figure 56. Photograph showing secondary Nitrogen Injection holes.

The experiments were carried out with this supersonic ejector nozzle under nominal cavity flow rates. The ejector nozzle is coupled to a mixing chamber with a divergence angle of 2°. The cold flow tests were performed, by passing 27.5mmol/s of nitrogen instead of oxygen through slits, without any flow through primary and secondary passages. The corresponding plenum pressure upstream of the oxygen passage was 12torr. However, when primary nitrogen flow of 370mmol/sec was also added in addition to the above without any flow through secondary, the plenum pressure increased to 24torr. This rise is basically explained by choking of singlet oxygen flow through slits by the primary flow streams. The cavity and pitot pressures measured inside laser cavity were 6torr and 61torr respectively. The addition of secondary nitrogen flow of 27.5 mmol/sec with earlier case resulted in cavity and pitot pressure of 7.5torr and 78torr respectively.

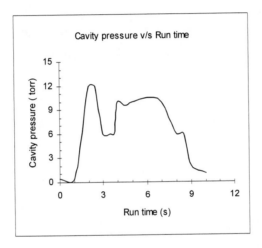

Figure 57. Temporal variation of cavity pressure.

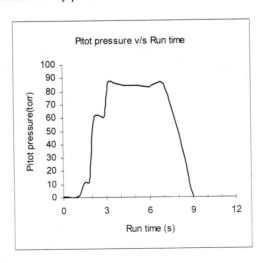

Figure 58. Temporal variation of Pitot pressure.

These results are in close agreement with the analytical results obtained using the aforesaid 1-D gas dynamic formulation. The cavity Mach number was found to drop marginally in case of hot runs to ~2.5 for a measured cavity pressure of ~10torr and Pitot pressure ~ 85torr. The cavity pressure is higher compared to that in the cold runs possibly due to the heat release and the accompanying boundary layer effects. In the present system, the heat addition in the cavity occurs due to the presence of a water vapor fraction of ~ 5%. Thus, the total pressure of the active media is expected to be nearly 170torr. Typically, for the optimal dilution ratio of 13:1 the variation of cavity and pitot pressures during the run are shown in Figs. (57) and (58).

It is clear from the figure that the dilution ratio has only a nominal effect on the cavity pressure and Mach number however the Pitot pressure decreases considerably. It is observed that conditions of high Mach number and high stagnation pressure occur at high dilution ratios, which are beneficial from both the pressure recovery and laser extraction point of view.

In addition to the directions indicated above, research in the area of COIL is also being pursued to generate pulses [49, 153-158], both through gain modulation as well as pulsed volume generation of the gain medium, second harmonic generation [159], mode locking [54] and output power stabilization [160.]

5.0. COIL Applications

5.1. Industrial Applications

On the industrial front, high power lasers are employed for various material processing applications like drilling, cutting, welding, surface hardening, alloying cladding, heat treatment etc. A detailed discussion on all these processes and review of various lasers for this kind of applications are given in ref [161]. Commercial high power industrial lasers must have qualities like low running cost, high average power, better beam quality, simple beam delivery system, capacity for long time run and ruggedness. The CO_2 and Nd.YAG lasers are the most commonly used in the industry since early sixties for material processing such as cutting, welding, cladding etc [162]. In the present day scenario, excimer, HF/DF and COIL are also being considered [163-165]. The analysis carried out by Bohn [166] indicate that as on date the CO_2 laser looks most attractive from the cost point of view, however it is worth noticing that the COIL wavelength is much shorter and its interaction with the materials is much stronger as compared to that of CO_2. Further, COIL wavelength is fiber compatible [167] and hence suitable for remote applications like underwater cutting etc. Though the cost of chemical fuels enhances the COIL running cost, however, it is believed that the regeneration/recycling concepts would bring down the running cost considerably. Even though the high power Nd.YAG lasers are fiber compatible and have already proved for remote operation , the average power of these lasers is in the range of 5 kW and its poor beam quality make these lasers inferior as compared to COIL.

COIL has excellent beam quality thus resulting in better brightness quality as compared to that of the other potential industrial lasers such as CO_2 or Nd. YAG lasers even at very high powers. Figure (59) representing the M^2 value of these lasers clearly shows that COIL has 3-4 times better beam quality as compared to the commercial CO_2 lasers at power levels of few tens of kW.

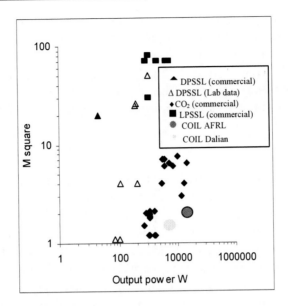

Figure 59. Beam quality factor survey [168].

COIL is a low pressure, low temperature system and the duration of the run is restricted mainly because of the degradation of the liquid fuel (BHP). In case, it is possible to recondition BHP on line by removing salt and extra water and replenishing the desired chemicals, COIL can run continuously day and night. Many workers have discussed these issues in detail [169, 170]. During prolonged runs, the reaction of BHP with chlorine gas results in the reduction of HO_2^- ion contents and thus degrades the performance of the COIL. Vetrovec [171] has suggested a practical scheme for regeneration of chemicals to bring down the running cost of systems based on COIL. In this proposed scheme, the regeneration takes place in combined electrolytic cells, which produce simultaneously the chlorine and BHP using commercial available technologies. The combined cell has two compartments one is for production of chlorine gas along with hydroxide (Brine electrolysis cell) and the other is for the BHP (Dow cell). The schematic of the closed loop BHP regeneration scheme is shown in Fig. (60)

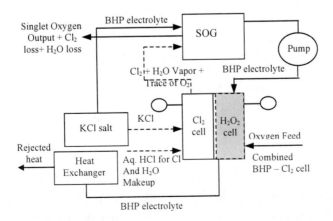

Figure 60. Configuration of the regenerative loop [171].

The BHP stream leaving the SOG is directed into the peroxide compartment (Dow cell) in which partially depleted HO_2^- ions are regenerated. The heat release in the Dow cell during this regeneration helps in dissolving the salt (KCl) and it remains in the solution. This regenerated BHP is chilled while passing through heat exchanger and during this process; the KCl gets precipitated and separated. This separated KCl is passed to the Brine cell compartment and through the electrolysis process chlorine gas is generated.

5.1.1. Cutting /Welding

Laser cutting is probably the earliest and the most established technique among the laser material processing mechanisms employing CW or repetitive pulsed high power lasers with power ranging from few hundreds of Watts to few kilo Watts. In laser cutting, focused laser beams heat, melt and vaporize the material along the cut and the molten metal or vapor from the kerf area is blown away by gas jets (called gas assisted cutting) simultaneously.

Yasuda et al. [165] and Kar et al. [172] have reported on the use of COIL for cutting of stainless steel. Carroll et al. [173], on the other hand, have used the 5 – 6 kW RADICL system not only for carbon steel but also for aluminum and have suggested detailed theory involved in the processes.

Various groups have modeled the processing capability of laser cutting and the interaction is illustrated in Fig (61).

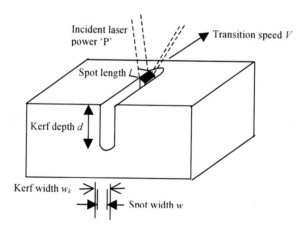

Figure 61. Typical Laser cut geometry.

Atsuta et al. have [174] developed an empirical model, generally known as Kawasaki model, using a lumped parameter technique as,

$$\frac{d}{P} = \frac{\alpha'}{V*w + \beta'} \qquad (117)$$

Here d/P is the scaled kerf depth, V*w describes the area processing speed, α' represents the absorptive property of the material at the incident laser wavelength and β' corresponds to the thermal properties of the material. This relation has been observed to explain results related to cutting of thin metal sheets. Scott et al. [175] have suggested a modified generalized model that considers the heat loss term but neglects the convection and radiation losses [175]. The relation is similar to that of the Kawasaki model and is given as,

$$\frac{d}{P} = \frac{\alpha}{V * w_k + \beta \sqrt{V * w_k}},$$

(118)

The main difference lies in definitions for α and β.

The schematic of the experimental setup used in RADICL is shown in Fig (62).

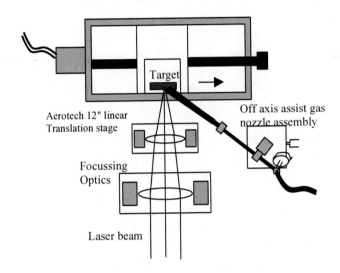

Figure 62. Schematic of the experimental setup used for cutting by RADICL COIL system [173].

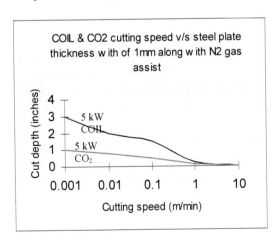

Figure 63. COIL and CO_2 cutting speed as the function of plate thickness and device power [177].

The RADICL system used 5-6 kW stable resonator power, focused to a spot size of 1.8 x 3 mm^2 using a combination of lenses. The system could demonstrate cut depths of 20 mm on aluminum targets using nitrogen gas assistance. On the other hand, it could cut carbon steel up to 41mm and 65 mm using nitrogen and oxygen assist gas respectively.

Further, it has been observed that COIL cuts carbon steel and stainless steel at the rates which are comparable to those achieved using Nd:YAG. However, these rates are approximately three times faster when compared to those obtained using CO_2 laser of same power [176]. Typical experimental cutting speed results using COIL and CO_2 lasers are shown in Fig. (63) [177].

Bohn [166] has demonstrated the possibility of using COIL laser for cutting concrete materials. Using a 10kW COIL, he could achieve a cutting depth of 5 cm with a Kerf width of about 1 mm at a cutting speed of about 1.2 m/minute. Extrapolation of these experimental studies indicate that a 70 kW COIL would be capable of cutting 30 cm thick concrete at a speed of 0.4m/minute.

In addition to efficient cutting applications, COIL can also be used for neat welding of materials [178]. The welding using laser, in general, is a costly affair but high production rate can offset this and high welding quality makes them cost effective. Laser welding is, however, superior because of its excellent features such as limited size of the heat affected zone, high welding speeds, narrow and deep processing, absence of filler material and applicability of the process in almost any atmosphere. In laser welding, focused region is heated to the melting point for sufficiently long duration to allow the melt to produce the joint. Quality of welding depends on the vaporization of the melt, size of the melt pool, thermal gradient etc.

COIL of 8.5 kW power has been used for welding a 10 mm thick stainless steel (304) plate at a speed of about 1.5m/minute [179]. Combination of 6kW pulse modulated Nd.YAG and 10kW continuous wave COIL system [180] has been developed for excellent quality welds. Full penetration welding with very high quality in SS304 plate thickness of 20 mm (in case of one sided welding) and 30 mm thickness (in case of double sided welding) has been demonstrated with a welding speed of 1m/minute. The observed welding depth for different welding speed for COIL, continuous wave Nd.YAG and pulsed wave Nd. YAG and their combination are shown in Fig. (64).

Based on the success of COIL assisted cutting and welding technologies, Carroll et al. [173] have demonstrated that the laser can also be used for cladding stainless steel substrates with ceramics. Laser cladding is effective in improving oxidation, corrosion and wear resistant properties of the materials by enhancing their microstructures. For their experiments stainless steel substrates of dimensions 87x49x6.5 mm^3 and the 2 mm thick mixture of ZrO_2 and Al cladding powder was pre-placed on 25x4 mm^2 area. The substrate was moved at a speed of about 0.5cm/s speed while keeping the laser spot fixed. The laser power on the target was about 4.7kW. The experiments with low Al content powder produced a crack free coating with fish scale like surface morphology while higher one showed a relatively smooth surface but with few cracks.

High power capability of COIL with its fiber compatibility makes them highly suitable for cutting or welding thick sections for under water applications. Under water cutting or drilling has wide applications such as maintenance of ships without taking them to dry dockyard, erection of huge structures in seawater as required for oil companies, rock crushing for oil well etc. Kawasaki Heavy Industries Japan has already demonstrated under water cutting on a 80mm thickness stainless steel work piece using 7kW COIL power through optical fiber [181].

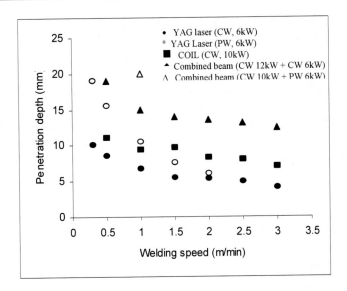

Figure 64. Penetration depth Vs welding speed for Nd: YAG, COIL and their combination [180].

5.1.2. Rock Crushing / Drilling

Under a Cooperative Research and Development Agreement (CRADA), the Colorado School of Mines demonstrated rock crushing with the COIL for oil-drilling applications [182]. Remote delivery of the beam has the advantage of removing the laser from the processing floor, thus addressing safety concerns associated with chemical laser operation while increasing the flexibility for cutting table layout to optimize production efficiency. This also allows for remote operations using robotic manipulators for material processing in hazardous environments or confined spaces. Because of the precise wavelength and controlled energy, COIL can eliminate problems with well control, sidetracks and directional drilling for gas wells at depths of more than 15,000 feet [182]. The lasers can cut through rock at high speeds as compared to those employing conventional techniques: typically laser can drill a gas well in 10 hours as compared to 10 days usually taken by conventional technique. Further, it can do so without affecting the outer edge of the rock the way drill bits do. A laser-drilling machine would vaporize rock thus avoiding drilling of mud and its associated problems.

5.1.3. Decontamination and Decommissioning Applications

The remote processing of materials for Decontamination and Decommissioning (D&D) of the obsolete nuclear reactors and nuclear material processing facilities is an important and difficult task. Survey shows that more than 500 such structures are required to be disposed worldwide in the coming 25 years span that may involve multi billion-dollar expenditure [167]. The conventional tools like oxy-acetylene torches, plasma arc cutters, band saws etc. have the draw backs like remote integrity, severe manual training, contamination of the equipments, risk of electrical shock, radiation hazards etc. Studies [167] indicates that lasers are most suitable for D&D applications because of their certain unique benefits such as fiber transportation and remote handling with suitable high power lasers, easy integration because of compactness and no tensions on target materials, fast processing because of the possibility of delivering very high power and cutting of complex and thick shapes/odd dimensions, minimal waste or debris material production, smallest kerf etc.

In addition, independent studies carried out by Oak Ridge National laboratory and the Idaho National Engineering Laboratory concludes that the laser technology can ultimately prove to be a cost effective technique in dismantling operations. High power lasers with average power of the order of 10kW, with remote handling capability, are required for the cutting of the thick walls of such nuclear systems. Also the same lasers can be used for the decontamination process by ablation technique. In the present day scenario, Nd.YAG and COIL are the only two lasers, which can be useful for D& D applications. But presently, the Nd. YAG maximum average power outputs are limited to about 5kW and thus COIL is a potential candidate for these applications as power scaling is not a problem.

The concept of D& D based on COIL system involves three separate vehicles: one each for fuel, main module and electronic control system. The laser beam from the laser cavity is transmitted through the optical fibers and robotic beam manipulator controls and carry out the dismantling operation.

5.1.4. Space Debris Removal

The space exploration, during the last few decades, has come across hundred thousands of space debris (of size 1 to 20 cm) in the low earth orbit which are serious threat and their removal is essential for future space missions. Acquiring and tracking these objects at the 100 to 200 km is a very difficult task. Phipps et al. [183] have reported that a repetitively pulsed CO_2 laser with average power 30 kW and 5 ns pulse width can effectively remove this debris. Pulsed lasers in the repetitive mode are preferred because of high peak power and better mechanical coupling efficiency [183]. Since COIL can be operated in the pulsed mode operation, [184] it can serve as one of the potential candidates for these applications. Bohn has estimated the COIL power requirements, for both pulsed and CW, for space debris removal applications. It has been concluded that about 1MW COIL peak power and a beam director of 6-7 m can solve the problem of debris removal existing up to altitudes of about 1000km [166].

The success of reliable COIL system development with regeneration capabilities and its adaptation for industrial applications has led Japanese scientists [110] to think about different modes of operating these systems. First mode is based on the realization of reliable multikilowatt (up to 20 kW) systems located at the industrial site where these can be conveniently used. These systems should be capable of running more than 24 hours continuously. The second mode of operation is mobile and requires a 30kW COIL system capable of running at least for two hours at a stretch. The mobile system can be conveniently taken to the place where industrial operations are required. One of the applications may include D & D. The third mode is based on the development of a 500 kW optical power station. The optical power station can run day and night and supply optical power through optical fibers to various industries.

5.2. Defense Applications

With the rapid development of high power lasers, Directed Energy Weapons based on lasers are now becoming a reality. These weapons based on HF / DF under the US-Israel joint

Natilus program have already shown their potential against Katyusha rockets [186]. DEW based on various lasers of different wavelength and power levels can address various defense applications such as, strategic and tactical missile defense, anti-satellite capabilities, air and ship defense, ground combat support etc. The applications of high power lasers in the area of defense and the ongoing worldwide programs have been nicely discussed and reviewed in ref [161]. The HPL based DEW has number of advantages which include speed of light delivery on to the targets, fast response/ reaction, high firing rate with minimum cost, no effect because of acceleration due to gravity, line of sight trajectory etc. However, effective use of laser based DEW systems requires thorough understanding of the physics and availabilities of technologies like source development with required energies and beam quality, sophisticated optics system for beam forming, transformation, tracking the target and pointing, knowledge / data of atmospheric conditions and adaptive optics control for compensating effects produced because of turbulence, knowledge / data of target material interaction.

In the present day scenario technology for COIL, HF/DF and CO_2: GDL are well mature and are thus the possible candidates as the DEW source. However with the fast development of heat capacity solid-state lasers, that has already reached a 15 kW average power levels [187], next decade would bring a tough competition between COIL and solid state lasers at least for ground based applications. Relative comparison of three mature lasers from the view point of DEW requirements, for tactical applications, is given in Table (3).

Table (3). A comparative assessment of three potential high power lasers

Property / Requirement	COIL	CO_2: GDL	DF	HF
Power scalability	Very high	High	Very high	Very High
Mass efficiency (kJ/Kg)	500	30-40	200	300
Compactness	Compact	Bulky	Bulky	Bulky
Beam quality	Best (M^2= 1.3) (θ=10 µrad)	Good (M^2= 3) (θ= 43 µrad)	Best (M^2= 2) (θ=25 µrad)	Best (M^2= 2) (θ=25 µrad)
Brightness on target	Best	Good	Better	Better
Atmospheric Transmission	Good	Better	Best	Good
Optics size	Smallest	Bigger	Smaller	Smaller
Beam pointing stability requirement	Less than 1 µrad	Typically 5 µrad	Typically 3 µrad	Typically 3 µrad
Effect on target	Best	Good	Better	Better
Adaptive optics	Most complicated	Less complicated	More complicated	More complicated
Operational easiness	Complicated	Less complicated	Complicated	Complicated
Safety	Low Toxic	Non Toxic	Highly Toxic	Highly Toxic

The above table clearly shows the superiority of COIL system for DEW applications. The countries like USA, Germany, Russia, China and Israel are already engaged in the development of high power COIL systems. USA's Air Borne Laser (ABL) and Airborne Tactical Laser (ATL) are notable systems and are in the process of being deployed in the near future.

5.2.1. Air Borne Laser (ABL)

The ABL project of USA started in 1996 is being realized jointly by three reputed organizations viz. (i) Boeing Defense Group for aircraft modification including battle management system (ii) TRW (Now taken by Northrop Grumman Space Technologies) for the megawatt class COIL source and (iii) Lockheed Martin Missile and Space for entire optical control. The artistic view of the ABL is shown in Fig. (65) [182]. It uses a modified commercial Boeing 747-400F aircraft, intended to fly at an altitude of about 40000 feet in order to destroy theatre ballistic missiles in their boost phase, from a distance of few hundred kilometers [188]. Initial plans were to use 12 COIL modules of 200 kW each in order to achieve 2.4 MW power for realizing the desired ranges. However, because of weight problems, it is now reported that in the first phase it would use only 6 modules and 1.2 MW of power and operate the system for reduced ranges. It is worth mentioning that the development of high efficient nozzle and supersonic mixing systems would soon overcome these issues and the initially planned system may be realized in the near future. In addition to the main COIL laser source, the ABL consists of six infrared sensors positioned on the fuselage which constantly scan all directions, either automatically or at the prompt of launch detecting satellites, for hot missile exhaust plumes. When these sensors see any missiles or series of missiles, the ABL's auxiliary lasers swing into action within seconds. Two solid-state lasers of different wavelengths (Nd: YAG and Yb: YAG of kW power level) illuminate the target continuously for tracking. A CO_2 laser is also used as an active ranger to provide continuously all the reference co-ordinates of the target with respect to ABL. The reflected

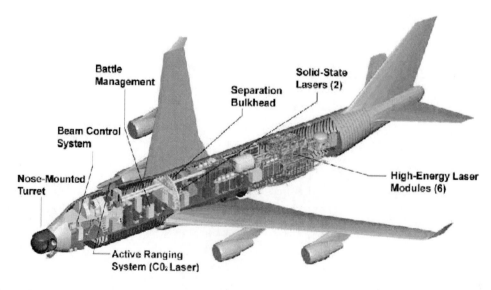

Figure 65. Artistic view of the ABL system [187].

beam energy from Nd:YAG laser is utilized to get the necessary atmospheric data for the feedback loop in adaptive optic system. Yb:YAG beam, on the other hand, tracks the image in real time and helps in locking onto the target. ABL is planed to carry sufficient chemicals for 20-40 missile shots with engagement duration of about 3 seconds.

The laser tests at Northrop Grumman have proved that the effectiveness of the laser to destroy a missile from stationary mounting [189]. All the six laser modules have successfully been fired for the first time in November 2004. The installation of beam /fire control system has been reported to be completed in Dec.2004. The COIL system is likely to be installed in the aircraft during late 2005 and the first prototype is expected to be ready in 2006.

5.2.2. *Advanced Tactical Laser (ATL)*

Advanced Tactical Lasers (ATL) of USA are planned to utilize a 200kW COIL system that will be mounted on a truck and would be aiming cruise missiles over 10 km range. The system is likely to have capability of shooting over 100 laser shots using a half-meter diameter beam director [186]. Rocketdynes division of Boeing USA has also planned for installing these COIL based ATLs on helicopters for deployment for both US Army and Navy applications. It is believed that US army is also considering a hundred kilowatt level COIL system mounted on a V-22 tilt rotor or CH-47 Chinook helicopter for its use for covert activities such as setting fires. All these systems are likely to have sealed exhaust system.

Further, the damaging of ineffective satellite with reduced power of MIRACL system has already been demonstrated [190]. Energy densities up to 10 W/cm^2 can be fatal for the satellite surfaces and can make them ineffective. Though these studies indicate vulnerability of satellites to lasers, nevertheless the solar panels based on new materials that are sensitive to 1.3-micron wavelength opens up new possibility of using COIL system for powering these satellites.

6. CONCLUSION

Chemical oxygen-iodine lasers have reached to a maturity level, where practical systems are now being configured around them. The technologies like regeneration, pressure recovery and efficient supersonic mixing would make them unique in the near future. Whereas, the development of variants such as Electri-COIL and All gas iodine laser would make this laser as the first choice in the next decade as well. As 21st century is going to be the era of laser based and microwave based high energy weapon, COIL based and heat capacity solid state laser based systems are likely to survive beyond 2020. Since electrical requirement for COIL systems except for Electri-COIL is minimal as compared to solid-state lasers, these lasers are going to be obvious choice for airborne and space applications. Even challenging applications such as debris removal and powering of satellites will become feasible in the next decade. Industrial applications of COIL have already been demonstrated. It is only on the basis of the initial successes and its superiority over other lasers, much larger systems are being planned. However, the development of efficient pressure recovery systems will help in making these systems compact and reliable.

REFERENCES

[1]   Kasper, J.V.V., Pimental, G.C. *App. Phys. Let.* 1964, 5, 231- 233.

[2]   McDermott, W.E., Pchelkin, N. R., Benard, D. J., Bonsek, R. R. *App. Phys. Let.* 1978,32, 469-470.

[3]   Demaria, A.J., Ultee, C.J. *App. Phys. Let.* 1966, 9, 67- 69.

[4]   McDermott, W.E., Stephens, J.C., Vetrovec, J., Dickerson, R.A. *SPIE,* 1997, 2987, 146- 156.

[5]   Endo, M., Osaka, T., Takeda, S. *App. Phys. Let.* 2004, 84, 2983- 2985.

[6]   Rosenwaks, S., Baramshenko, B.D., Bruins, E., Furman, D., Rybalkin, V., Katz, A. *SPIE,* 2001, 4631, 23-33.

[7]   http://www.boeing.com/defense-space/military/abl/, 2005.

[8]   Tei, K., Sugimoto, D., Endo, M., Takeda, S., Fujikova, T. *SPIE,* 1999, 3887, 162 -169.

[9]   Zagidullin, M.V., Nikolaev, V.D. *SPIE,* 1999, 3688, 54-61.

[10] Zagidullin, M.V., Nikolaev, V.D., Svistun, M.I., Khvatov, N.A. *Quantum Electronics,* 1998, 28, 400 – 402.

[11] Elior, A., Lebiush, E., Schall, W.O., Rosenwaks, S. *Optics and Laser Technology,* 1994,26, 87 - 89.

[12] Balyvas, I., Barmashenko, B.D., Furman, D., Rosenwaks, S., Zagidullin, Z. *IEEE Journal of Quantum Electronics,* 1996, 32, 2051- 2057.

[13] Rittenhouse, T.L., Phipps, S.P., Helms, C.A., Truesdell, K.A. *SPIE,* 1996, 2702, 333- 338.

[14] Zagidullin, M.V., Nikolaev, V.D., Svistun, M.I., Khvatov, N.A., Ufimtsev, N.I. *Quantum Electronics,* 1997,27, 195-199.

[15] Fujji H., Yoshida, S., Iizuka, M., Atsuta, T. *J. App. Phys.,* 1990, 67, 3948-53.

[16] Monroe, D.K. *SPIE,* 1994, 2121, 276- 283.

[17] Endo, M., Tei, K., Sugimoto, D., Nanri, K., Uchiyama, T., Fujioka, T. *SPIE,* 2001, 4184, 23 - 26.

[18] Carroll, D.L., Verdeyen, J.T., King, D.M., Zimmerman, J.W., Laystorm, J.K., Woodard, B.S., Benavides, G.F., Kittell, K.W., Solomon, W.C. *IEEE quantum electronics,* 2005, 41, 213-223.

[19] Henmshaw, T.L., Manke II, G.C., Madden, T.J., Berman, M.R., Hager, G.D. *Chemical Physics Letters,* 2000, 325, 537-544.

[20] Yuryshev, N.N. *Quantum Electronics,* 1996, 26, 567-584.

[21] Richardson, R.J., Wiswall, C.E., Carr, P.A.G., Hovis, F.E., Lilenfeld, H.V. *J. App. Phys.,* 1981, 52, 4962– 4969.

[22] Hon, J. *AIAA Journal,* 1996, 34, 1595 - 1603.

[23] Tyagi, R.K., Rajesh, R., Singhal, G., Mainuddin, Dawar, A.L., Endo, M. *Journal of Infra Red Physics and Technology,* 2003, 44, 271-279.

[24] Endo, M., Nagatomo, S., Takeda, S., Zagidullin, Z., Nikolaev, V.D., Fujii, H., Wani, F., Sugimoto, D., Sunako, K., Nanri, K., Fujioka, T. *IEEE Journal of Quantum Electronics,* 1998, 34, 393-398.

[25] Furman, D., Bruins, E., Barmashenko, B.D., Rosenwaks, S. *App. Phys. Let.* 1999, 74, 3093-3095.

[26] Barmashenko, B.D., Rosenwaks, S. *AIAA Journal,* 1996, 34, 2569 - 2574.

[27] Hashimoto, T., Nakano, S., Hachijin, M., Komatsu, K., Mine, Y., Hara, H. *Applied Optics*, 1993, 32, 5936 - 5943.

[28] Barmashenko, B.D., Rosenwaks, S. *Applied Optics*, 1996, 35, 7091- 7101.

[29] Cohen, L.S., Coulter, L.J., Egan, W.J. *AIAA Journal*, 1971, 9, 718 - 724.

[30] Kodymova, J., Spalek, O., Jirasek, V. *AIAA Paper* 2000- 2426, 2000.

[31] Bruins, E., Furman, D., Rybalkin, V., Barmashenko, B.D., Rosenwaks, S. *IEEE Journal of Quantum Electronics*, 2002, 38, 345-352.

[32] Trozzolo, A.M. Editor. *Ann, NY Acad. Sci.* 1970 171, 1-302.

[33] Kenner, R.D., Khan, A.U. *Chem. Phys. Letters*, 1975, 36, 643- 646.

[34] Turro, J. *Modern molecular Photo Chemistry*, Benjamin Illuming Publications, California, 1978.

[35] Derwent, R.G., Thrush, B.A. *Chemical Physics Letters*, 1971, 9, 591-592.

[36] Khan, A.U., Kasha, M. *Journal of Chemical Physics*, 1963, 39, 2105- 2106.

[37] Kanofsky, R., *Journal of American Chemical Society*, 1986, 108, 2977-2979.

[38] Caminade, J.A.M., Khatib, F.L., Koenig, M., Aurby, J.M. *Canadian Journal of Chemistry*, 1985, 63, 3203-3209.

[39] Kanofsky, J.R. *Journal of Organic Chemistry*, 1986, 51, 3386-3388.

[40] Benard, D.J. *J. of Physical Chemistry*, 1996, 100, 8316-8322.

[41] Benard, D.J., McDermott, W. E., Pchelkin, N.R., Bousek, R.R. App. *Phys. Let.* 1979, 34, 40 – 41.

[42] Truesdell, K.A., Helms, C.A., Hager, G.D. *SPIE*, 1995, 2502, 217 - 237.

[43] Kalinovsky, V.S., Kirillov, G.A., Konovalov, V.V., Nikolaev, V.D. *SPIE*, 1993, 1980, 138 – 147.

[44] Rosenwaks, S., Bacher, *J. App. Phys. Let.* 1982, 41, 16-18.

[45] Watanabe, K., Kashiwabara, S., Sawai, K., Toshima, S., Fujinoto, R. *IEEE Journal of Quantum Electronics*, 1983, 19, 1699-1703.

[46] Zhang, R., Chen, F., Song, X., Xu, Q., Huan, C. *Chinese Journal of Lasers*, 1988, 15, 455-457.

[47] Yoshida, S., Endo, M., Sawano, T., Amano, S., Fuji, H. *J. App. Phys.*, 1989, 65, 870-872.

[48] Zolotarev, V. A., Kryukov, P.G., Podmarkov, Y.P.,Frolov, M.P., Yuryshev, N.N. *Quantum Electronics*, 1989, 19, 709 – 710.

[49] VanMarter, T. Heaven, M.C., Plummer, D. *Chemical Physics Letters*, 1996, 260, 201-207.

[50] Zagidullin, M. V., Kurov, A.Y., Kuprianov, N.L., Nikolaev, V.D., Svistun, M.I., Erasov, N.Y. *Quantum electronics*, 1991, 21, 747 - 753.

[51] Schmiedberger. J., Kodymova, J., Spalek, O., Kovar, J. *IEEE Journal of Quantum Electronics*, 1991, 27, 1265-1270.

[52] McDermott, W. E., Stephens, J.C., Vetrovac, J., Dickerson, R.A. *SPIE*, 1997, 2987, 146-156.

[53] Carroll, D.L., Solomon, W.C. *SPIE*, 2001, 4184, 40 – 44.

[54] Phipps, S.P., Helms, C.A., Copeland, R.J., Rudolph, W., Truesdell, K.A., Hagger, G.D. *IEEE Journal of Quantum Electronics*, 1996, 32, 2045-50.

[55] Entress- Fursteneck, L.V., Handke, J., Gruenewald, K.M., Bohn, W.L., Schall, W.O. *SPIE*, 1997, 3092, 706-709.

[56] Zagidullin, M.V., Nikolaev, V.D., Svistun M.I., Khvatov, N.A. *Quantum Electronics*, 2000, 30, 161 - 166.

[57] Nikolaev, V.D., Zagidullin, M.V., Hager, G.D., Madden, T.J. *AIAA Paper* 2000- 2427, 2000.

[58] Yang, T.T., Bhowmik, A., Burde, D.H., Clark, R., Carroll, S., Dickerson, R.A., Eblen, J., Gylys, V.T., Hsia, Y.C., Humphreys, R.H., Moon, L.F., Hurlock, S.C., Tomassian, A. *SPIE*, 4760, 2002, 537-549.

[59] Tyagi, R.K., Rajesh, R., Singhal, G., Mainddin, Dawar, A.L., Endo, M. *Optics and Laser Technology*, 2003, 35, 395-399.

[60] Belousova, V.P., Belousova, I.M., Grenishin, A.S., Danilov, O.B., Kislev, V.M., Kris'ko, A.V., Mak, A.A., Murav'eva, T.D., Sosnov, V.N. *Optics and Spectroscopy*, 2003, 95, 830-832.

[61] Manke II, G.C., Cooper, C.B., Dass, S.C., Madden T.J., Hager, G.D. *IEEE J. Quantum Electronics*, 2003, 39, 995-1001.

[62] Carroll, D.L., Verdeyen, J.T., King, D.M., Zimmerman, J.W., Laystorm, J.K., Woodard, B.S., Benavides, G.F., Kittell, K., Stafford, D.S., Kushner, M.J., Solomon, W.C. *App. Phys. Let.* 2005, 86,111104,

[63] Spalek, O., Kodymova, J., Hirsi, A. *J. App. Phys.*, 1987, 67, 2208- 2211.

[64] Yoshida, S., Fujii, H., Sawano, T., Amano, S., Fujioka, T. *App. Phys. Let.*, 1987, 51, 1490 - 1492 .

[65] Yoshida, S., Saito, H., Fujikova, T., Yamakoshi, H., Uchiyama, T. *App. Phys. Let.* 1986, 49, 1143 - 1144.

[66] Avizonis, P.V., Truesdell, K.A. *SPIE*, 1995, 2502, 180 –200.

[67] Schall, W.O., O. S., Duschek, F.R. SPIE, 1997, 3092, 686 -689.

[68] Blauer, J.B., Munjee, S.A., Truesdell, K.A., Curtis, E.C., Sullivan, J.F. *J. App. Phys.*, 1987, 67, 2508 - 2517.

[69] Wani, F., Endo, M., Vyskubenko, B.A., Ilyin, S.P., Kurkovsky, I.M., Takeda, S., Fujioka, T. *IEEE J. Quantum Electronics*, 1998, 34, 2130 - 2137.

[70] Adamenkov, A., Vyskubenko, B.A., Il'in, S.P., Krukovski, I.M. *Quantum Electronics*, 2002, 32, 490 – 494.

[71] Muto, S., Kawano, T., Endo, M., Nanri, K., Taeda, S., Fujioka, T. *SPIE*, 2001, 4184, 91-94.

[72] Balan, N.F., Gizatullin, R.M., Zagidullin, Z., Kurov, A.Y., Nikolaev, V.D., Pichkasov, V.M., Svistun, M.I. *Quantum Electronics*, 1989, 19, 1412-1415.

[73] Zagidullin, M.V., Nikolaev, V.D., Svistun, M., Ufimtsev, N.I. *SPIE*, 1996, 2767, 221- 228.

[74] Kodymova, J., Spalek, O. *Japanese Journal of Applied Physics*, 1998, 37, 117 - 121.

[75] Zagidullin, M.V. *SPIE*, 1998, 3574, 569 -576.

[76] Shimizu, K., Sawano, T., Tokuda, T., Yoshida, S., Tanaka, I. *J. App. Phys.*, 1991, 69, 79-83.

[77] Spalek, O., Kodymova, J., Hirsl, A. *SPIE*, 1988, 1031, 319-323.

[78] Mallik, A., Dawar, A.L., Murthy, Y.L.N., Kohli, K., Badola, A.K., Razdan, A.K. *SPIE*, 1999, 3612, 102 - 110.

[79] Helms, C.A. *SPIE*, 2001, 4184, 13 - 18.

[80] Yang, B., Zhuang, Q., Sang, F., Chen, F., Zhang, C. *AIAA Paper* 95 – 1931, 1995.

[81] Schall, W.O., Kraft, D. *SPIE*, 1996, 2767, 229 -236.

[82] Zagidullin, M. *SPIE*, 1995, 2502, 208 - 216.

[83] Barmashenko, B.D., Furman, D., Rosenwaks, S. *SPIE*, 1998, 3574, 273 -280.

[84] Barmashenko, B.D., Elior, A., Lebiush, E., Rosenwaks, S. *J. App. Phys.*, 1994, 75, 7653 – 7665.

[85] Carroll, D.L., King, D.M., Fockler, L., Stromberg, D., Solomon, W.C., Sentman, L.H., Fisher, C.H. *IEEE J. Quantum Electronics,* 2000, 36, 40 - 51.

[86] Rajesh. R., Hussain. M., Zaidi. Z. H., Tyagi. R.K., Singhal, G., Mainuddin, Dawar. A.L. Endo. M. *Journal of Infrared and Millimeter Waves*, 2004, 25, 2361-2382.

[87] Spalek, O., Kodymova, J., Zagidullin, M., Nikolaev, V.D. *SPIE*, 1997, 3092, 565 - 568.

[88] Tyagi, R.K., Rajesh, R., Singhal, G., Mainuddin, Dawar, A.L., Endo, M. *SPIE*, 2003, 4971, 11 – 21.

[89] Liu, W.F., Chen, F., Han, X., Sang, F. *SPIE*, 2001, 4184, 128 - 131.

[90] Stilvast, W.T. *Laser Fundamentals*, ISBN 81-7596-040-X, Cambridge University Press, Cambridge, U.K., , First South Asian paperback Edition, 1998, p329-65.

[91] Fujii, H., Atsuta, T. *SPIE*, 1992, 1980, 148-152.

[92] Fujji, H., Atsuta, T. *SPIE*, 1997, 3092, 700-705.

[93] Rigrod, W.W. *J. App. Phys.*, 1963, 14, 2602-2609.

[94] Yang, B. *SPIE*, 1998, 3574, 281-289.

[95] Sang, F., Yang, B., Zhuang, Q. *SPIE*, 2001, 4184, 27 - 31.

[96] Boreisho, A.S., Mal'kov, V.M., Savin, A.V., Vasil'ev, D.N., Evdokimov, I.M., Trilis, A.V., Strakhov, S.Y. *Quantum Electronics*, 2003, 33, 307-311.

[97] Yoshida, S., Shimizu, K., Tahil, H., Tanaka, I. *IEEE J. Quantum Electronics,* 1994, 30, 160 - 166.

[98] Endo, M., Kawakami, M., Takeda, S., Nanri, K., Fujioka, T. *SPIE*, 1999, 3612, 62 - 70.

[99] Shapiro, A.H. *The dynamics and thermodynamics of compressible fluid flow* ISBN 0 471-06691-5, John Wiley & Sons, New York, Volume I, 1953, pp 73-154

[100] Azyazov, V.N., Zagidullin, M.V., Nikolaev, V.D., Safonov, V.S. *Quantum Electronics*, 1997, 27, 477 - 480.

[101] Crown, J.C. " *Supersonic Nozzle Design"*, NACA, Technical Note No. 1651, 1948.

[102] Heidner , R.F., Gardner, C.E., Segal, G.I., El-Sayed, T.M. *J. Phys. Chem.*, 1983, 87, 2348 - 2360.

[103] Yang, T.T., Cover, R.A., Quan, V., Smith, D.M., Bauer, A.H., McDermott, W.E., Copeland, D.A. *SPIE*, 1997, 2989, 126 - 149.

[104] Carroll, D.L. *AIAA Journal*, 1995, 33, 1454 -1462.

[105] Miller, J.H., Shang, J.S., Tomaro, R.F., Strang, W.Z. *J. Propulsion and Power,* 2001, 17, 836 - 844.

[106] Carroll, D.L. *AIAA Journal*, 1996, 34, 338 – 346.

[107] Eppard, W.M., McGory, W.D., Godfrey, A.G., Cliff, E.M., Borggaard, J.T. *AIAA Paper* 2000-2576, 2000.

[108] Walter, R.F., O'Leary, R.A. *SPIE*, 1992, 1980, 206-212.

[109] Acebal, R. *SPIE*, 1994, 2119, 59-67.

[110] Endo, M., Sugimoto, D., Takeda, S., Nanri, K., Fujioka, T. *SPIE*, 2000, 3889, 438 - 446.

[111] Yang, T.T., Hsia, Y.C., Moon, L.F., Dickerson, R.A. SPIE, 2000, 3931, 116-130.

[112] Yang, T.T., Dickerson, R.A., Moon, L.F., Hsia, Y.C. *AIAA Paper*, 2000–2425, 2000.

[113] Rybalkin, V., Katz, A., Barmashenko, B.D., Rosenwaks, S. *App. Phys. Let.* 2003, 82, 3838 – 3840.

[114] Sugimoto, D., Wani, F., Endo, M., Takeda, S., Fujioka, T. *AIAA Paper* 99-3426, 1999.

[115] Waitz, I.A., Qui, Y.J., Manning, T.A., Fung, A.K.S., Elliot, J.K., Kerwin, J.M., Kransodebski, M.N., Sullivan, O., Tew, D.E., Greitzer, E.M., Marble, F.E., Tan, T.S., Tillman, T.G. *Prog. Aerospace Science*, 1997, 33, 323 - 351.

[116] Hirata, T., Endo, M., Shinoda, K., Osaka, T., Nanri, K., Takeda, S., Fujioka, T. *SPIE*, 2003, 5120, 410 - 419.

[117] Endo, M., Sugimoto, D., Okamoto, H., Takeda, S., Fujioka, T. Japanese *J. App. Physics*, 2000, 39, 468-474.

[118] Endo, M., Kawakami, M., Takeda, S., Wani, F., Fujioka, T. *SPIE*, 1999, 3612, 56 - 61.

[119] Mikheyev, P.A., Shepelenko, A.A., Voronov, A.I., Kupryaev, N.V. *Quantum Electronics*, 2002, 32, 1-4.

[120] Spalek, O., Censky´, M., Jira´sek, V., Kodymova´, J., Jakubec, I., Hager, G.D. *IEEE J. Quantum Electron*, 2004, 40, 564 - 570.

[121] Wakazono, T., Hashimoto, K., Takemoto, T., Uchiyama, T., Muro, M. *SPIE*, 1998, 3574, 290-294.

[122] Jirasek, V., Spalek, O., Kodymova, J. *SPIE*, 2001, 4184, 103 – 106.

[123] Spalek, O., Jirasek, V., Kodymova, J., Censky, M., Jakubek, I. *SPIE*, 2002, 4631, 34-42.

[124] Hewett, K.B., Hager, G.D., Crowell, P.G. *Chemical Physics*, 2005, 308, 159–169.

[125] Grimley, A.J., Huston, P.L. *Journal of Chemical Physics*, 1978, 69, 2339-2346.

[126] Gylys, V.T., Yang, T.T. *Proceedings of the International conference on Lasers*, 99, San Diego, A, 1991, 163-168.

[127] Vetrovec, J., Yang, T.T., Copeland, D.A. *SPIE*, 2000, 3931, 71 – 80.

[128] Zalesskii, V.Yu. *Sov. Physics JETP*, 1975, 40, 14-17.

[129] Hill, A, *Proceedings of International conference on Lasers* 2000, Albuquerque, USA.

[130] Schmiedberger, J., Fujji, H. *SPIE*, 1995, 2502, 338 –343.

[131] Fujji, H., Itami, S., Kihara, Y., Fujisaki, K., Okamura, M., Yoshitani, E., Yano, K., Miyatake. T., Schmiedberger, J. *SPIE*, 2000, 4065, 818 - 825.

[132] Schmiedberger, J., Tikahashi, S., Fujji, H. *SPIE*, 1997, 3092, 694 - 697.

[133] Napartovich, A.P., Deryugin, A.A., Kochetov, I.V. J. Physics D: *Applied Physics*, 2001. 34, 1827-1833.

[134] Ionin, A.A., *J. Physics D: Applied Physics*, 2003, 36, 982-989.

[135] Rawlins, W.T., Lee, S., Kessler, W.J., Davis, S.J. *App. Phys. Let.* 2005, 86, 051105.

[136] Zalesskii, V. Yu. *Quantum Electronics*, 1983, 13, 701 – 707.

[137] Belousova, I.M., Danilov, O.B., Mak, A.A., Belousov, V.P., Zalesskii, V.Yu., Grigor'ev, A.A., Kris'ko, A.V., Sosnov, E.N. *Optics and Spectroscopy*, 2001, 90 , 858-862.

[138] Belousova , I.M., Danilov, O.B., Mak, A.A., Belousov, V.P., Zalesski, V.Y., Grigorev, V.A., Krisko, A.V., Sosnov, E.N. *Optics & Spectroscopy*, 2001, 90, 773- 777.

[139] Danilov, O.B., Belousova, I.M., Mak, A.A., Zalessky, V.Y., Grigorev, V.A., Krisko, A.V., Sosnov, E.N., Belousov, V.P. *SPIE*, 2001, 4351, 92 - 98.

[140] Kodymova, J., Spalek, O., Rohlena, K., Beranek, J. *SPIE*, 1996, 2702, 320 – 332.

[141] Boreysho, A.S., Khailov, V.M., Mal'Kov, V.M., Savin, A.V. *SPIE*, 2001, 4134, 401 - 405.

[142] Haefer, R.A. *Cryopumping Theory and Practice*, Chapter 5, Claredon Press, Oxford, UK, 1989.

[143] *Union Carbide's Molecular Sieves*, product brochure of UOP, Inc., Des Plaines, IL, USA.

[144] Welch, K.M. "*Capture Pump Technology*," Chapter 5, Pergamon Press, New York, NY, 1991

[145] Stern, S.A., DiPaolo, F.S. *J. Vacuum Sc. and Tech.*, 1969, 6, 941 - 950.

[146] Vetrovec, J. *SPIE*, 2000, 3931, 60 - 70.

[147] Fabri, J., Paulon, J., *NACA* –1410, 1953.

[148] Arbel, A., Shklyar, A., Hershgal, D., Barak, M., Sokolov, M. *Journal of Fluids Engineering, ASME*, 2003, 125, 121 - 129.

[149] Emanuel, G. *AIAA Journal*, 1976, 14, 1124 - 1125.

[150] Hook, D., Magiawala, K., Haflinger, D., Behrens, H. *AIAA paper* 92-2981, 1992.

[151] Keenan, J.H., Neumann, E.P., Lustwerk, F. J. *Applied Mechanics*, 1950, 17, 299 - 309.

[152] Nikolaev, V.D., Zagidullin, M.V., Svistun, M.I., Anderson, B.T., Tate, R.F., Hager, G.D. *IEEE J. Quatum Electronics*, 2002, 38, 421 – 428.

[153] Schimiedberger, J., Kodymova, J., Kovar, J., Spalek, O., Trenda, P. *IEEE J. Quantum Electronics*, 1991, 27, 1262-1264.

[154] Schimiedberger, J., Spalek, O., Kodymova, J., Kovar, J. *SPIE*, 1992, 1810, 521-524.

[155] Kodymova, J., Spalek, O., Schmiedberger, J. *SPIE*, 1995, 2502, 304-309

[156] Hager, G.D., Kopf, D., Hunt, B.S., Anderson, B., Woolhiser, C., Crowell, P. *IEEE J. Quantum Electronics* , 1993, 29, 933-943.

[157] Basov, N. G., Vagin, N.P., Kryukov, P.G., Nurligareev, D.K., Pazyuk, V.S., Yuryshev, N.N. *Quantum Electronics*, 1984, 14, 1275 – 1276.

[158] Pazyuk, V.S., Vagin, N.P., Yuryshev, N.N. *SPIE*, 1996, 2767, 206-208.

[159] Hager, G.D., Hanes, S.A., Dreger, M.A. *IEEE J. Quantum Electronics*, 1992, 28, 2573-2576.

[160] Sugimoto, D., Kawano, T., Endo, M., Takeda, S., Fujioka, T. *SPIE*, 1998, 3574, 295 – 300.

[161] Premasundaran, M. *High power lasers and their applications*, Law and Commercial publications, 2004, New Delhi, India, I[st] edition, p 105-190.

[162] Belforte, D., Levitt, M. *The Industrial Laser Annual Handbook*, 1990 Edition, Tulsa, Oklahoma: Penn Well Books, 1990.

[163] Joeckle, R., Gautier, B., Nett, J., Schellhorn, M., Sontag, A., Stern, G. *AIAA paper* 95-1921, 1995

[164] Vetrovec, J., Hindy. R., Subbaraman. G. Spiegel, L. *SPIE*, 1996, 3092, 780-788.

[165] Yasuda, K., Atsuta, T., Sakurai, T., Okado, H., Hayakawa, H., Adachi, J. 3rd *JSME/ASME Joint International conference* on Nuclear Engineering, 1995, 1769.

[166] Bohn, W.L. *SPIE*, 2002, 4631, 53 - 59.

[167] Vetrovec, J., Hallada, M., Seiffert, S., Walter, O. *ICALEO* '99, SanDiego, CA, 1999, paper no. 1408.

[168] S. Shah, *Rocketdyne/ETEC document No.* 25599, US DOE, 1998.

[169] Fujji, H., Yoshida, S., Iizuka, M., Astute, T. *J. App. Phys.*, 1989, 66, 1033-1037.

[170] Truesdell, K.A., Lonergan, T., Wisniewski, C., Healey, K. Scott, J., Helms, C. *AIAA paper* 94-2441, 1994.

[171] Vetrovec, J. *SPIE*, 1997, 2987, 157-165.

[172] Kar, A., Scott, J.E., Latham, W.P. *Journal of Laser Applications,* 1996, 8, 125 - 131.

[173] Carroll, D.L., Rothenflue, J.A., Kar, A., Latham, W.P. *Proceedings of the ICALEO* -96, Michigan, USA, 1996.

[174] Atsuta T., Yasuda, K., Matsumoto, T., Sakurai, T., Okado, H. *Proceedings of conference on Lasers and Electro-Optics (CLEO),* 1994.

[175] Scott, J.E., Rothanflue, J.A., Latham, W.T., Kar, A. *SPIE,* 1996, 2702, 339 - 346.

[176] Carroll, D.L., Rothanflue, J.A. *Journal of Laser Applications,* 1997, 9, 119 - 123.

[177] Carroll, D.L., King, D.M., Fockler, L., Stromberg, D., Madden, T.J., Solomon, W.C., Sentman, L.H., Fisher, C.H. *AIAA paper* 98-2992, 1998.

[178] Yasuda, K., Muro, M. *Review of Laser Engineering,* 2000, 28, 755 - 762.

[179] Nakabayashi T, and Muro, M. *SPIE,* 2000, 3888, 359 - 366.

[180] Wani, F., Nakabayashi, T., Hayakawa, A., Suzuki, S., Yasuda, K. *SPIE,* 2002, 4631, 128 - 136.

[181] Okado H, Sakurai T, Adachi J, Miyao H and Hara K, *SPIE,* 2000, 3887, 152-160.

[182] Article by Dr. William P. Latham of the Air Force Research Laboratory, USA published in site http://www.afrlhorizons.com

[183] Phipps, C.R., Reilly, J.P. *SPIE,* 1997, 3092, 728-731.

[184] Monrec, D.K. *SPIE,* 1994, 2121, 276 - 262.

[185] Hager, G.D., Kopf, D. 25th *AIAA Plasma dynamics and laser conference,* Colorado Springs Co, 1994.

[186] Mowthorpe, M. *"The Revolution in Military Affairs and Directed Energy Weapons",* Air & Space power chronicles –8th March 2002.

[187] *Directed energy weapons, An ATS Analysis & Discussion,* www.atsnn.com/story/36894.html, 8th *March 2004.*

[188] Forden, G.E. *IEEE Spectrum,* 1997, 34, 40-49.

[189] Article by R. A. Hettinga, 2003, http://popsci.printthis.clickability.com.

[190] Frank Sietzen, "Laser Hits Orbiting Satellite in Beam Test", *Space Daily,* October 20, 1997 [accessed] http://www.spacedaily.com/news/laser-97a.html.

In: Perspectives in Optics Research
Editor: Jeffrey M. Ringer, pp. 171-192

ISBN: 978-1-61122-934-9
© 2011 Nova Science Publishers, Inc.

Chapter 6

CHALCONES: POSSIBLE NEW MATERIALS FOR THIRD-ORDER NONLINEAR OPTICS

A. John Kiran[a], K. Chandrasekharan[*b],
Balakrishna Kalluraya[c] and H. D. Shashikala[a]

[a] Department of Physics, National Institute of Technology Karnataka,
Surathkal, Mangalore, India
[b] Lasers and Nonlinear Optics Lab, Department of Science and Humanities,
National Institute of Technology Calicut, Kerala, India
[c] Department of Chemistry, Mangalore University, Karnataka, India

ABSTRACT

'Search for potential materials' for third-order nonlinear optics has been of continuing interest in recent years. In this context, organic molecules are increasingly being recognized as the materials of the future because their molecular nature combined with the versatility of synthetic chemistry can be used to alter and optimize molecular structure to maximize third order nonlinear response. Chalcones have received considerable interest as materials for second-order nonlinear optical applications due to their ability to crystallize in noncentrosymmetric structure and their blue light transmittance. Being charge transfer compounds, chalcones can also possess large third-order nonlinearities due to their π-conjugated structure. In this article, we discuss the structure-nonlinear response relationship among a few chalcones and the possibility of using them for third-order applications. A meager or no work, to our knowledge, has been done so far on these molecules in this regard. Z-scan and degenerate four-wave mixing techniques were employed to investigate third-order optical nonlinearities of chalcone derivatives. Some of these molecules possess large $\chi^{(3)}$ of magnitude as high as 10^{-12} esu and exhibit strong optical limiting properties. Possible mechanisms that are responsible for optical limiting property of these molecules have been discussed.

[*] Corresponding Author: - E-mail: csk@nitc.ac.in; Phone: +91-495 2287343.

INTRODUCTION

Nonlinear optics (NLO) is attracting increasing attention around the world because of its applications in telecommunications, all optical switching, sensor protections, and possibilities for optical information storage and coupling [1, 2]. To build devices for such applications, it is necessary to identify either existing materials or new materials that possess desired nonlinear optical properties. Materials for nonlinear optics have developed considerably in recent years, with novel semiconductors, organo-metallic complexes, composite glasses embedded with metal nanoparticles, and organic polymers and molecules attracting much attention [3-7]. Among these materials, organic molecular materials, because of their unique chemical structures (π bonding), exhibit the largest nonresonant (nonabsorptive) optical nonlinearities [8, 9]. The strong delocalization of electrons in these materials determines a very high molecular polarizability and thus remarkable third order nonlinearities. The most promising application of nonlinear optical activity is optical limiting, i.e. the increase of optical absorption with light intensity. Materials possessing large two-photon and reverse saturable absorptivities are being investigated for their nonlinear absorption based optical limiting applications [10].

Chalcones are 1,3-Diphenylpropenone (benzylideneacetophenone) and its derivatives formed by substitution: ArCH=CHC(=O)Ar, as shown in figure 1.

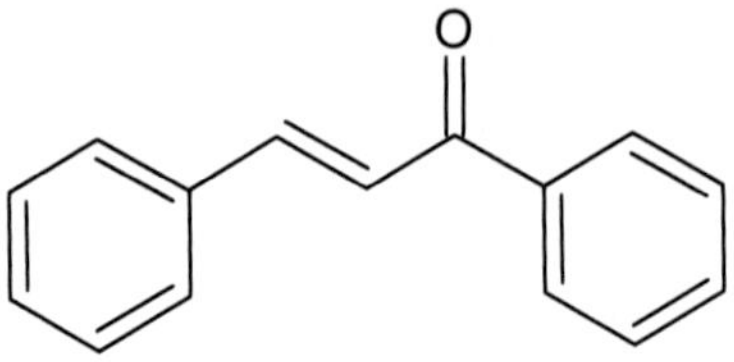

Figure 1. General structure of chalcones: ArCH=CHC(=O)Ar.

Chalcones are low molecular weight charge transfer (CT) compounds that show large nonlinear optical (NLO) properties without a longer wavelength absorption, as found in π-electron conjugated polymers. Charge transfer compounds have a large nonlinear electronic polarization that is caused by a large dipole moment change from a ground state to an excited state by optical radiation [11]. The molecular structure of intramolecular CT compounds distorts the distribution of π electrons in such a way that it results in the enhancement of optical nonlinearity. Such CT compounds are expected to possess advantages such as high solubility in organic solvents and matrix polymers and ease of sample preparation, as well as the possibility of giving high $\chi^{(3)}$ value and hence usefulness in photonic applications [11].

Till today, several CT compounds and CT complexes have been studied for their large third order nonlinearities [12-15]. By selecting an appropriate π-conjugation length and chemical structures of donor and acceptor, several new organic molecular crystals such as stilbene [13] and azobenzene derivatives [11], which exhibit $\chi^{(3)}$ larger than 10^{-12} esu, have been developed. The $\chi^{(3)}$ of 4-(N,N-diethylamino)-4'-nitrostilbene (DAENS), an intramolecular CT compound, has been reported in a solution, in a polymer matrix, and in a crystalline state [16, 17] and the maximum value obtained is 6×10^{-12} esu for real $\chi^{(3)}$ and 4×10^{-12} esu for imaginary $\chi^{(3)}$ [17]. The NLO properties of a novel CT compound, 4-(N, N-diethylamino)- 4'-nitrostyrene (DEANST), have been investigated by means of an optical Kerr shutter experiment, and a $\chi^{(3)}$ value of 7.4×10^{-12} esu has been measured in DEANST crystals [18, 19]. Some novel symmetrical short π-conjugated molecules, like SBA and SBAC, with large $\chi^{(3)}$ have been developed [20]. SBA and SBAC are benzylidene aniline derivatives that are symmetrically substituted by donor groups at both ends of the π-conjugations and have nitrogen atoms in their π-conjugation systems. In the off-resonant regions, the $\chi^{(3)}$ value in these molecules is measured to be as high as 4×10^{-11} esu [20].

Chalcones have been extensively studied for their second order nonlinear optical properties in the past few years [21]. However, being charge transfer compounds, chalcones can also possess large third-order nonlinearities due to their π-conjugated structure. Further, unlike second order nonlinearity, the advantage that the third order optical nonlinearity does not require any symmetry restrictions, has widen the range of possible structures of chalcone derivatives for consideration for third order nonlinear optical applications.

Various chalcones like, 1-(4-methylphenyl)-3-(4-methoxyphenyl)-2-propen-1-one, [22] 1-(4-chlorolphenyl)-3-(4-chlorolphenyl)-2-propen-1-one [23], 1,5-diphenylpenta-2, 4-dien-1-ones [24], 4-Br-4'-methoxy chalcone single crystals [25], arylfluranylpropenones [26], 1-(4-methylphenyl)-3-(phenyl)-2-propen-1-one [27], and 1-(4-methylphenyl)-3-(4-N, N dimethylaminophenyl)-2-propen-1-one [28, 29] have been studied for their second order NLO properties by several researchers. Chalcones have emerged as promising materials for second harmonic generation (SHG). Presently, green to blue wavelength lasers are required for high-density optical memories, color displays, and so on. Laser diodes and laser diode-pumped solid-state lasers have several advantages such as compactness, low cost, and long life. The harmonic generation of these lasers is very promising to obtain blue-green lights. As output powers of laser diodes are low, optical materials with large nonlinear susceptibilities are necessary to obtain efficient SHG. Organic crystals with conjugated π electrons are attractive candidates because of the large nonlinear optical coefficients. Among these organic crystals, the candidates must satisfy the following conditions: 1) they should be a certain size in bulk, 2) they should be transparent in the blue-green region and 3) they should be mechanically hard and chemically stable [29]. Chalcone derivatives are one of the candidates satisfying the above conditions because they have large nonlinear optical coefficients and a relatively short cutoff wavelength of transmission. Despite being recognized for second order applications, chalcone derivatives can also possess large third order nonlinearities. However, so far to our knowledge, a meager or no studies on the third order nonlinear optical (TNLO) properties of chalcones have been reported.

Apart from these optical properties, chalcones show some other interesting properties related to color reactions and biological activities. They show halochromism [30-32]. When they are wetted with concentrated H_2SO_4 they develop halochromic colors. When the intensely colored solution of chalcone in concentrated H_2SO_4 is treated with a little concentrated HNO_3, characteristic color change occurs [33]. The differing halochromic effects produced with concentrated H_2SO_4 serve to characteristic chalcones. Trans-chalcones usually show two absorption bands located at 300 nm (band I) and 330 nm (band II), of which band I results from the conjugation of the whole molecule [34]. The presence of enone function in the chalcone molecule confers antibiotic activity (bacteriostatic/ bactericidal) upon it. This property is enhanced when substitution is made at the α (nitro- and bromo-) and β- (bromo- and hydroxylic-) positions. Some substituted chalcones and their derivatives including some of their heterocyclic analogues have been reported to possess some interesting biological properties that are detrimental to the growth of microbes, tubercle bacilli, malerial parasites and intestine worms. Some of the compounds are claimed to be toxic to animals and insects and are also reported to exhibit inhibitory action on several enzymes, fungi and herbaceous plants. Compounds of the chalcone series also show a profound influence on the cardiovascular, cerebrovascular and neuromuscular systems [35-37].

We investigated some chalcones like, Dibenzylideneacetone (DBA) and derivatives, and 1-3-diaryl-propenones containing 4-methylthiophenyl moieties. We observed a good third order nonlinear response in all molecules, and interestingly a variation of the response among them that could be well associated with their molecular structure. A strong delocalization of π-electrons in these conjugated systems determines a very high molecular polarizability and hence remarkable third order nonlinearities [2]. Although great efforts have been expended on the investigation of the relationships that exist between the molecular structure of an organic material and the resulting third order nonlinearities, the complete understanding of these dependences is still incomplete. However, recent research efforts have enhanced the theoretical understanding and have thus improved processing techniques [38-40]. The study of the linear and nonlinear optical coefficients is fundamental for increasing the ability to tune nonlinear optical properties by appropriate design of organic systems at the molecular level.

Dibenzylideneacetone (DBA) has one more importance as a very good precursor for the preparation of many types of heterocyclic compounds of medical and biological importance.

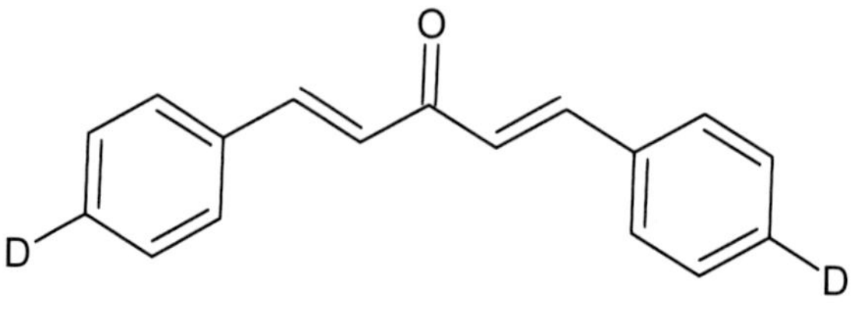

Figure 2. Structure of DBA. D is OCH3, Cl, N(CH3)2 for p-methoxyDBA, p-chloroDBA and p-(N,N-dimethyl)DBA respectively.

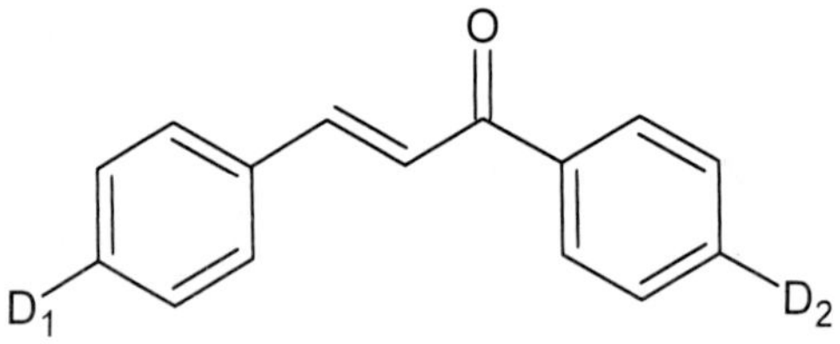

Figure 3. Structure of 1-3-diaryl-propenones containing 4-methyl thiophenyl moieties coded as mc-4, mc-6, mc-10, and mc-3. D_1 is H_3CS, common to all the four compounds. D_2 is CH_3, CH_3O, C $(CH_3)_3$, and Cl respectively for mc-4, mc-6, mc-10, and mc-3.

Practically, molecular design can play an important role in the development of third order materials by using the tremendous flexibility a molecular material offer to modify its chemical structure. Marius Alabota et al. have proposed criteria to enhance the third order nonlinearity in molecular materials [41]. According to their design criteria, the third-order nonlinearity of conjugated organic compounds can be enhanced by (i) increasing the conjugation length, to increase the distance over which charge can be transferred; (ii) creating a donor-acceptor-donor motif by substitution to increase the extent of charge transfer from the ends of the molecule to the center and (iii) reversing the sense of symmetric charge transfer by substituting electron acceptors and donors, and there by creating acceptor-donor-acceptor compounds . Similar strategy was used to design these chalcones through derivatization by substituting proper electron donors and electron acceptors and thus to create donor-π-acceptor-π-donor structures. The structures of compounds are given figure 2 and 3 below. Synthesis and detailed TNLO characterization of these compounds has been reported by John Kiran et al. [42, 43].

EXPERIMENT

Two experiments, a single-beam Z-scan and Degenerate Four Wave Mixing (DFWM) were employed to study TNLO properties of chalcones in their solutions in Dimethylformamide (DMF) (1×10^{-2} mol/L). DFWM was used in order to compare $\chi^{(3)}$ values of molecules with those $\chi^{(3)}$ obtained through Z-scan technique. A detailed theories and description of experiments are available in the literature [44, 45]. Z-scan is a simple and well-known technique used to measure nonlinear refractive index (n_2) and nonlinear absorption coefficient (β) of a material. The technique is particularly useful when the nonlinear refraction is accompanied by nonlinear absorption and it allows the simultaneous measurement of n_2 and β. The technique relies on the phenomenon of self-focusing or self-defocusing of an optical beam by a thin medium.

Four-wave mixing refers to the interaction of four waves in a nonlinear medium via the third order polarization. It results from the intensity dependent refractive index. When all the waves have same frequency, though their wave vectors are different, it is called as degenerate four wave mixing. It is one of the most important and versatile techniques, which provide information about the magnitude and response of the third order nonlinearity. There are several geometries used in studying this phenomenon, for example, forward geometry, backward geometry or phase conjugate geometry, nearly degenerate four wave mixing, and folded boxcar geometry [45]. However, the principle for all these geometries is the following: two beams interfere to form some type of grating and a third beam scatters off this grating, generating the fourth beam named the conjugate or signal beam. We used the standard phase conjugate geometry in our experiment.

The linear absorption spectra of DBA and its derivatives in DMF recorded with a UV-VIS Ocean Optics (Model SD 2000) spectrophotometer are depicted in figure 4. It can be seen that the excitation wavelength (532 nm), except for p-(N, N-dimethyl)DBA, is far away from the resonance wavelength, indicating that we measure only the nonresonant nonlinearity. This is true in all the samples, except for p-(N, N-dimethyl)DBA, investigated in this chapter. Linear refractive indices of samples in DMF were obtained through an Abbe refractometer.

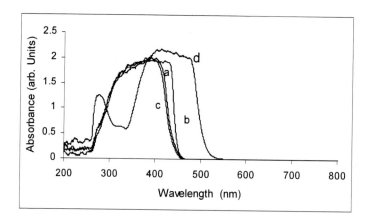

Figure 4. Linear absorption spectra of compounds dissolved in DMF (1×10^{-2} mol/L). a: DBA; b: p-methoxyDBA; c: p-chloroDBA; d: p-(N, N-dimethyl)DBA.

RESULTS

The single beam Z-scan with Nd: YAG nanosecond (7 ns) laser was performed on these chalcones dissolved in DMF. Details of experiment can be found in reference [42]. Both closed aperture and open aperture scans were performed in order to extract information on nonlinear refraction (NLR) and nonlinear absorption (NLA) of the samples in DMF solutions. Z-scan curves obtained for dibenzylideneacetone are shown in figure 5, 6 and 7 below. Solid lines in the figure are theoretical fits of experimental data. Details of equations used for curve fitting can be found in reference [44, 46]. The normalized transmittance for open aperture curve is given by

$$T(z) = \frac{\ln[1 + q_0(z)]}{q_0(z)} \text{ for } |q_0(z)| < 1 \tag{1}$$

where $q_0(z) = \frac{\beta I_0 L_{eff}}{(1 + z^2 / z_0^2)}$, z is the position of sample with respect to focal plane of the lens, z_0 is the Rayleigh range of the lens and $L_{eff} = (1 - e^{-\alpha L}) / \alpha$ is the length of the sample. The normalized transmittance for closed aperture scan is given by

$$T(z) = 1 + \frac{4\Delta\phi_0 x}{(1 + x^2)(9 + x^2)} - \frac{2\Delta\varphi_0(3 + x^2)}{(1 + x^2)(9 + x^2)} \tag{2}$$

where $\Delta\phi_0 = \dfrac{\Delta T_{p-v}}{0.406(1-S)^{0.25}}$ for $|\Delta\phi_0| \leq \pi$ and $\Delta\varphi_0 = \dfrac{1}{2}\beta I_0 L_{eff}$ are phase shifts due to NLR and NLA respectively [47].

In a closed aperture Z-scan curve, both nonlinear refraction and nonlinear absorption will be present; in which case, a simple division of the closed aperture Z-scan data by the one with open aperture is followed to obtain a pure nonlinear refraction curve. In this case the normalized transmittance would become [44]

$$T(z) = 1 - \frac{4\Delta\phi_0 x}{(1+x^2)(9+x^2)} \qquad (3)$$

Second Order Hyperpolarizability

The third order nonlinear susceptibility is related to the second order hyperpolarizability, which is a microscopic parameter. Second order hyperpolarizability γ_h of a molecule in an isotropic medium is related to the third order bulk susceptibility as below [48]:

$$\gamma_h = \chi^{(3)} / L^4 N. \qquad (4)$$

In the above equation, N is the density of molecules in the unit of molecules per cm^3. The term L is the local field factor, which in the Lorentz approximation is given by

$$L = \left(\frac{(n^2+2)}{3} \right) \qquad (5)$$

Here n is the linear refractive index of the medium.

Closed aperture curve demonstrates a strong negative refractive nonlinearity of the sample. n_2 was found to be as high as -3.512×10^{-12} esu. The values of real and imaginary parts of $\chi^{(3)}$ were found to be -0.38×10^{-13} esu and 0.10×10^{-13} esu respectively. Similar results were observed with derivatives of DBA. The values of $\chi^{(3)}$ are tabulated in table 1. It has been shown by J. Kiran et al., [42] that upon substitution of strong electron donor groups to DBA, nonlinearity can be enhanced according to the design criteria discussed by M. Alabota et al. [41]. Interestingly it has been found that the third order nonlinearity in these chalcones varies according to the π electron density in the molecules.

The values of n_2, Re $\chi^{(3)}$ and Im $\chi^{(3)}$ reported in table 1 were obtained by repeating the Z-scan on each sample, and the values were found to be consistent in all trials with an error of less than 10 %. The increased nonlinear response in the case of p-methoxyDBA is mainly due to the enhanced π – electron density in the molecule [41, 49].

The results show that the increase in $\chi^{(3)}$ of p-methoxyDBA is nearly double compared to the $\chi^{(3)}$ value of DBA. Comparing to chlorine, which is an electron-withdrawing group due to high electronegativity, methoxy is an electron releasing by resonance. So, among these compounds the order of π-electron density is p-methoxyDBA > DBA > p-chloroDBA > o-chloroDBA.

Among p-chloroDBA and o-chloroDBA although both are having chlorine substituents, in the case of o-chloroDBA, chlorine is in very close vicinity of double bond, so it withdraws electrons more powerfully than chlorine present in the para -position. Consequently, o-chloroDBA showed least nonlinear response among all other compounds. It showed no detectable nonlinear response at the input energy used during our experiment. Further, to determine the contributions of the solvent to n_2 we conducted Z-scan experiment on the pure DMF, and found that neither nonlinear refraction nor nonlinear absorption was observed at the input energy used. Hence, any contribution from solvent to the nonlinearity of the samples is negligible. We did not observe any significant scattering of laser beam from sample solutions within the energy limit used during our experiments.

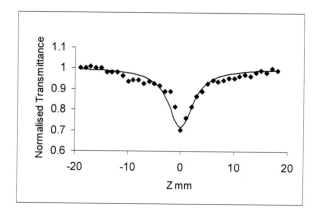

Figure 5. Open aperture curve for DBA. Solid line is a theoretical fit of experimental data with β = 0.72 cm/GW to equation (1).

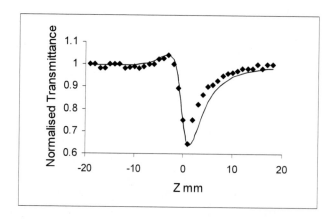

Figure 6. Closed aperture curve for DBA. Solid line is a fit of experimental data with $\Delta\Phi_0$ = 0.97 and β = 0.72 cm/GW to equation (2).

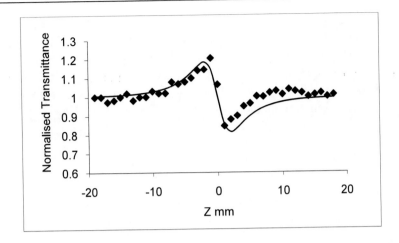

Figure 7. Pure NLR curve for DBA obtained through division method. Solid line is a fit of experimental data with $\Delta\Phi_0 = 0.97$ to equation (3).

Table 1. Values of n_2, β, γ_h, real and imaginary $\chi^{(3)}$ determined experimentally for different DBA based compounds

Name	n_0	n_2 esu	β cm/GW	$\gamma_h \times 10^{-32}$ esu	Re $\chi^{(3)}$ $\times 10^{-13}$ esu	Im $\chi^{(3)}$ $\times 10^{-13}$ esu
DBA	1.425	-3.512×10^{-12}	0.72	0.2	-0.38	0.10
p-chloroDBA	1.412	-3.18×10^{-12}	0.98	0.184	-0.33	0.11
p-methoxyDBA	1.417	-5.745×10^{-12}	0.77	0.325	-0.613	0.12
o-chloro DBA	1.413	No detectable response				

A similar variation of nonlinear response was observed with 1-3-diaryl-propenones containing 4-methyl thiophenyl moieties, whose structure is shown in figure 3. John Kiran et al. [43] have done detailed study on these molecules. These molecules exhibit strong negative refractive nonlinearity and two-photon absorption at 532 nm. The nonlinear response was found to increase on going from mc-3 to mc-10. Among different electron donors substituted in these compounds, the electron donating ability of C $(CH_3)_3$ > OCH_3 > CH_3 > Cl. As a result, the nonlinear response is found to be the highest in the case of mc-10. In mc-10 there is an S-CH_3 group on the left hand side of the molecule and a C $(CH_3)_3$ on the right hand side. The S-CH_3 group is known to be a good electron donor and it gives away electrons easily to form a stable π-electron distributed system. This donates more electrons into the molecule and on the other side C $(CH_3)_3$ also releases electrons in to the molecule. The acceptor, C=O, at the center accepts electrons and hence there is a strong delocalization of electrons in the molecule. There is a symmetric charge transfer from the ends to the center and hence the extent of π -electron transfer is increased in the molecule. In mc-3, mc-4 and mc-6 the response was observed to vary according to the density of delocalized electrons in the molecule. Therefore, the enhanced nonlinear response of mc-10 is mainly due to the increased π-electron density in the molecule [41]. In mc-3, due to the inductive effect of chlorine, the π-electron density and the charge transfer in the system is reduced considerably. Therefore the response is minimum compared to that of other three samples.

In order to get more information on the structure-NLO property relationships, three more similar molecules were synthesized. Their structure is shown in figure 8 below. These molecules possess one Br atom in their structure.

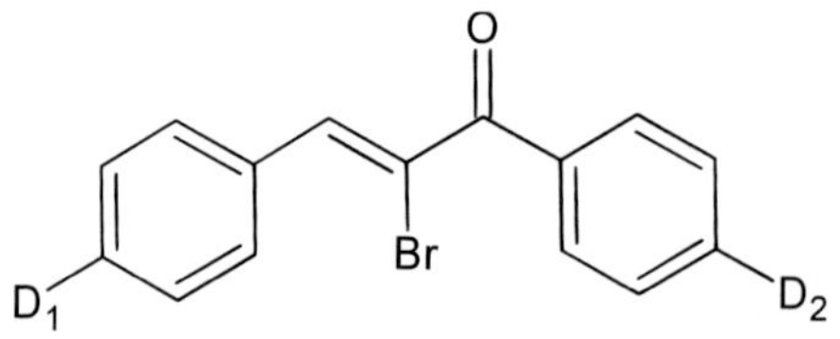

Figure 8. The structure of compounds coded mc-3m, mc-4m and mc-6m. D_1 is H_3CS common to all compounds. D_2 is Cl, CH_3 and OCH_3 respectively for mc-3m, mc-4m and mc-6m.

The nonlinear repose of molecules was found to decrease upon substitution of Br atom. The observed decrease in NLO response of mc-3m to mc-6m may be explained as follows. In compounds mc-3m, mc-4m and mc-6m there is a Bromine atom at the center of molecule. Br is a good electron acceptor due to its high electonegativity. Bromine, being a heavy atom, affects the planarity of the molecule, which in turn reduces the delocalization of electrons through out the molecule. In conjugated systems, like those of chalcones, co-planarity is required for the transfer of electrons from donor to acceptor. In α-Bromo chalcones, the bulky Bromine atom sterically interferes with adjacent cis- phenyl ring and pushes this phenyl ring out of plane. Hence there is no co-planar arrangement of two phenyl rings with the double bond and the π- conjugation is thus not effective in the transfer of electrons from one end to the other in the system [50].

Table 2. Experimentally determined values of n_0, n_2, β, $γ_h$, Re $\chi^{(3)}$ and Im $\chi^{(3)}$ for different Compounds

Name of the sample	n_0	n_2 10^{-11} esu	β cm/GW	$γ_h$ $\times 10^{-32}$ esu	Re $\chi^{(3)}$ 10^{-13} esu	Im $\chi^{(3)}$ 10^{-13} esu
mc-3	1.423	-1.098	1.472	0.616	-1.181	0.23
mc-4	1.421	-1.21	2.046	0.681	-1.296	0.26
mc-6	1.421	-1.23	2.690	0.931	-1.556	0.42
mc-10	1.420	-2.023	3.035	1.147	-2.170	0.47
mc-3m	1.423	-0.719	1.440	0.415	-0.770	0.22
mc-4m	1.421	-0.251	0.330	0.141	-0.269	0.05
mc-6m	1.423	-0.33	0.321	0.184	-0.356	0.049

As a result, nonlinear response of these samples was found to decrease upon substitution of Br atom at position –2 of the propenones. The dependence of third-order NLO properties on donor and acceptor type of substituents clearly demonstrates the electronic origin of nonlinearity of these compounds, and hence the possibility of dominant contribution of thermal effects to the observed nonlinearity can be ruled out.

The values of determined NLO coefficients of mc-3 to mc-10 are given in table 2. These values of third order optical parameters were found to be consistent in all repeated scans with an error of less than 12 %.

Nonlinear Absorption

All compounds discussed so far are transparent at 532 nm but exhibit strong two-photon absorption (TPA). Generally, nonlinear absorption (NLA) can be caused by free carrier absorption, saturated absorption, direct multiphoton absorption, or excited state absorption. However, if the mechanism belongs to a simple two-photon absorption, β should be a constant that is independent of the on-axis irradiance I_0 [51]. Further, from basic theoretical consideration, it is seen that the TPA-induced decrease of transmissivity can be expressed as [52].

$$T(I_0) = \frac{I(L)}{I_0} = \frac{1}{1 + I_0 L \beta} \; , \tag{4}$$

where I_0 is the on-axis intensity of the beam within the sample of thickness L. If the laser beam has a gausian transverse intensity distribution in the medium, then the TPA-induced transmissivity change will be

$$T(I_0) = \frac{I(L)}{I_0} = \frac{\ln(1 + I_0 L \beta)}{I_0 L \beta} \tag{5}$$

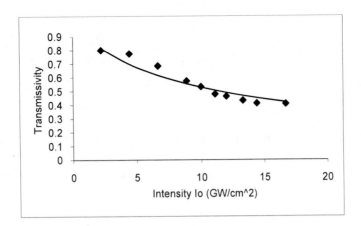

Figure 9. Transmissivity versus input intensity for p-methoxyDBA in DMF solution (1×10^{-2} mol/L). Solid line is a fit of data to equation (5) with a best-fitting parameter of $\beta = 2.2$ cm/GW.

Hence, if the transmissivity of a sample varies as in equation (5) we may assume that the dominant mechanism for nonlinear absorption is TPA. Therefore to confirm TPA we plotted transmissivity measured as a function of input intensity for p-methoxyDBA, as shown in figure 9 and a good correlation was seen between the experimental data and the theoretical fit of equation (5) [53]. Figure 10 shows a plot of β versus input intensity I_0 for p-methoxyDBA and mc-10, where it is clear that β is almost independent of I_0. A similar behavior was observed with other samples also. This confirms that the dominant mechanism of nonlinear absorption in our samples is a TPA-induced absorption.

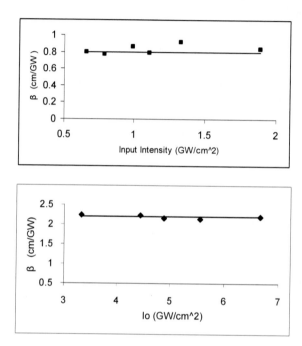

Figure 10. Plots of β versus I_0 for a) p-methoxyDBA and b) mc-10. β is nearly a constant independent of I_0.

Optical Limiting

Optical limiting is a nonlinear optical (NLO) process in which the transmittance of a material decreases with increased incident light intensity. Optical limiters have been utilized in a variety of circumstances where a decreasing transmission with increasing excitation is desirable. However, one of the most important applications is eye and sensor protection in optical systems [10]. These devices, which work due to intrinsic properties of the materials used for their fabrication, are used to control the amplitude of high intensity optical pulses. An ideal optical limiter exhibits a linear transmission below threshold, but above its threshold its output intensity is constant. For a material without significant NLO absorption (linear material) the transmitted light intensity will increase proportionally to the incoming light intensity (i.e. the relative transmission is constant). For a material with nonlinear optical absorption the transmitted intensity will increase proportionally to the incoming light intensity while the intensity is low. However, when a certain input intensity is reached the NLO absorption will start and the increase in the outgoing light will be reduced for increasing light intensities (i.e. the relative transmission decreases with increased intensity). For some materials this optical limiting effect can be so strong that the transmitted light intensity reaches a constant value independent of the incoming light intensity. This is defined as the clamping level. However, pulses with realistic temporal and spatial profile modify this response. Under realistic conditions, the limiting threshold is less well defined, and the output fluence will not be perfectly clamped at a constant value.

Based on the strong two-photon absorption, good optical limiting of nanosecond laser pulses can be expected from these chalcones. Figures 11 and 12 below show optical limiting behavior of p-methoxyDBA and some mc-series compounds respectively.

In the case of p-methoxyDBA, for input energies less than 300 μJ/pulse, the output energy increases linearly with incident energy. However, in excess of 300 μJ/pulse the output energy is almost clamped at 230 μJ/pulse. The nonlinear absorption was observed to increase on going from mc-3 to mc-10. In the case of mc-10, for input energies less than 220 μJ/pulse, the output energy increased linearly with the incident energy. But for energies more than 220 μJ/pulse, the output energy was almost constant assuming the value near 140 μJ/pulse. The limiting threshold increased from 220 μJ/pulse to 250 μJ/pulse in the case of mc-4 and the output was clamped at around 153 μJ/pulse. This threshold further increased to 370 μJ/pulse in the case of mc-3 and there the output became almost constant at 210 - 220 μJ/pulse. Thus, optical limiting behavior of these samples was found to vary according to the π-electron delocalization in molecules.

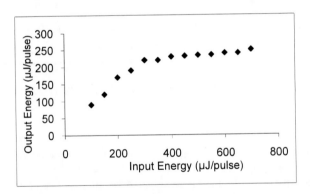

Figure 11. Optical limiting behavior of p-methoxyDBA.

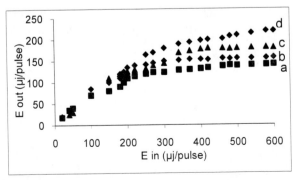

Figure 12. Optical limiting of nanosecond pulses in samples a=mc-10, b=mc-4, c=mc-6 and d=mc-3.

p-(N,N-dimethyl)Dibenzylideneacetone

It is one of the derivatives of DBA. It has been shown that the nonlinear response increased upon substitution of N, N-dimethyl group to DBA and this compound possesses the highest nonlinearity among other DBA derivatives [42]. Though electron-donating ability of N, N-dimethyl group is high compared to methoxy group and hence compound shows high NLO response, careful studies show that the compound has an absorption tail at 532 nm. This small linear absorption can also contribute to the observed large nonlinearity of the sample. Therefore, further study was done on this sample to explore possible mechanism responsible for its large nonlinearity. The linear absorption coefficient α of sample in DMF (1×10^{-2} mol/L) at 532 nm was found to be 0.023 cm^{-1}. The Z-scan [with a 50% (S=0.5) aperture] was repeated on this solution at pulse energy of 16 μJ, which corresponds to a peak irradiance of 0.355 GW/cm^2. The Re $\chi^{(3)}$ value of 1.9×10^{-12} esu was obtained for p-(N,N-dimethyl)DBA.

The $\chi^{(3)}$ value of the compound is very high and is a little more compared to 1.3×10^{-12} esu, obtained for tricyanovinyl azo dye and comparable with 6×10^{-12} esu, obtained for 4-(N,N-diethylamino)-4'-nitrostilbene (DEANS) [16]. This shows that the compound can be a promising dopant material for doping polymers for optical limiting and other photonic applications.

Since the compound absorbs at 532 nm, there could be some contribution from excited state absorption. The excited state cross section σ_{ex} can be measured from the normalized open aperture Z-scan data [54]. We assume that the molecular energy levels can be reduced to a three level system in order to calculate σ_{ex}. Molecules are optically excited from the ground state to the singlet-excited state. The molecules from this state relax either to the ground state or the triplet state, when excited state absorption can occur from the triplet to the higher triplet excited state.

The change in the intensity of the beam as it passes through the material is given by $\frac{dI}{dz} = -\alpha I - \sigma_{ex} N(t) I$, where I the intensity, and N is the molecules in the excited state. The excited state density of molecules appears as a result of a nonlinear absorption process whose intensity dependence can be obtained from $\frac{dn}{dz} = \frac{\sigma_{ex} I}{h\nu}$, where ν is the frequency of the laser. Combining the above two equations and solving for the fluence of the laser and over the spatial extent of the beam gives the normalized transmission for open aperture as

$$T = \ln\left(1 + \frac{q_0}{1+x^2}\right) / \frac{q_0}{1+x^2} \qquad (6)$$

where $q_0 = \frac{\sigma_{ex}\alpha F_0 L_{eff}}{2h\nu}$, F_0 is the fluence of the laser at the focus and $L_{eff} = \frac{(1 - exp^{-\alpha L})}{\alpha}$.

A fit of equation (6) to the open aperture data at 532 nm with q_0, yields $\sigma_{ex} = 6.24 \times 10^{-17}$ cm^2 for the sample solution. The ground state absorption cross-section of sample calculated from $\alpha = \sigma_g N_a C$, where N_a is Avogadro's number and C is the concentration in moles/cm^3, is found to be $\sigma_g = 3.818 \times 10^{-18}$ cm^2. The value of σ_{ex} is larger than the value of σ_g, which is in agreement with the condition for observing reverse saturation absorption [54, 55]. Reverse saturation absorption generally arises in a molecular system when the excited state absorption cross section is larger than the ground state cross section. If the absorption cross section of excited state is smaller than that of ground state, then the material becomes more transparent and bleaches i.e. it is a saturable absorber. If the mechanism belongs to the simple two-photon absorption, β should be a constant that is independent of the on-axis irradiance I_0. As if the mechanism is the direct three-photon absorption, β should be a linear increasing function of I_0 and the intercepts on the vertical axis should be nonzero [51]. But figure 13 shows that β is decreasing with increasing I_0. The fall-off of β with increasing I_0 is a consequence of the reverse saturable absorption [55]. Hence, the mechanism responsible for strong NLA of the compound was found to be reverse saturation absorption.

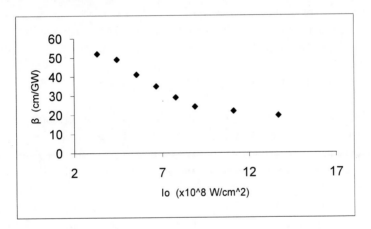

Figure 13. A fall-off of β with increase in on axis intensity within the sample I_0.

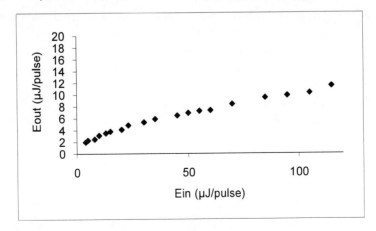

Figure 14. Optical limiting of nanosecond pulses in p-(N,N-dimethyl)DBA in DMF.

Optical limiting based on the reverse saturable absorption can be expected for the nanosecond laser pulses. Figure 14 shows strong optical limiting behavior of the compound solution. For input energies less than 15 μJ/pulse, the output energy increased linearly with the incident energy. But for energies more than 15 μJ/pulse, the measured output energy deviates from linearity, indicating the occurrence optical limiting. The linear transmittance of this sample at 532 nm was 70%.

Doping Polymers

Despite these compounds possess very high third order nonlinearity, they cannot be directly used in practical devices since they can get degraded when exposed to intense laser beams. One can overcome this problem and make effective use of these materials in devices by doping them into a polymer matrix, as this can enhance the opto-chemical and opto-physical stability as well as mechanical and thermal properties, while retaining the NLO properties and linear optical transparency [56]. Here we report on doping one of these compounds, p-(N,N-dimethyl)DBA, into poly(methylmethacrylate) (PMMA). PMMA has been selected as a matrix because it is a hard, rigid but transparent polymer with glass transition temperature of $125°$ C, and an average molecular weight of 60,000. PMMA has physical durability which is far superior to that of other thermoplastics and it is tougher than polystyrene [57].

PMMA and p-(N,N-dimethyl)DBA were taken in the powder form and dissolved in DMF. The concentration of the dopant in PMMA matrix was varied from 2.5 % to 15% [sample I: 2.5%; sample II: 5%; sample III: 10%; sample IV: 15%]. The Z-scan was obtained with a 50% (S=0.5) aperture and at pulse energy of 16 μJ, which corresponds to a peak irradiance of 0.355 GW/cm^2. In order to avoid cumulative thermal effects, data were collected in single shot mode [58]. Further, to determine the contributions of the solvent and PMMA to the observed NLO properties, we conducted Z-scan experiment on the pure DMF and PMMA dissolved in DMF. Neither nonlinear refraction nor nonlinear absorption was observed at the input energy used. Hence, any contribution from the solvent and PMMA to nonlinearity of the sample is negligible. Therefore, the nonlinear response observed was solely due to p-(N,N-dimethyl)DBA. Z-scan results obtained are given in table 3 below. It can be seen that response is increasing with concentration of the sample in PMMA. However, not much increase was observed with sample IV, where the linear transmittance became very less (~ 40 %). Figure 15 shows the optical limiting behavior of p-(N,N-dimethyl)DBA doped PMMA.

Table 3. Calculated values of β, n_2, Re $\chi^{(3)}$, and Im $\chi^{(3)}$ for doped PMMA

Sample	β cm/GW	$n_2 \times 10^{-10}$ esu	Re $\chi^{(3)} \times 10^{-12}$ esu	Im $\chi^{(3)} \times 10^{-12}$ esu
Sample I	48.62	1.375	1.479	0.752
Sample II	52.90	1.552	1.602	0.801
Sample III	57.50	1.700	1.829	0.890
Sample IV	59.60	1.770	1.886	0.886

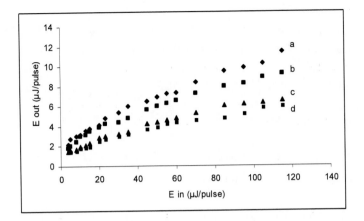

Figure 15. Optical limiting of nanosecond pulses in p-(N,N-dimethyl)DBA doped PMMA at different concentrations. a: sample I; b: sample II; c: sample III; d: sample IV.

This study shows that p-(N,N-dimethyl)DBA is an attractive optical limiting material and can be used for doping polymers for optical limiting applications.

Concentration Dependence of Nonlinearity

Concentration dependence of NLO coefficients can be analyzed to extract information on the NLO properties of the solute. The concentration of solutes in the solution was varied and Z-scan measurements were repeated on solutions at each concentration to study the variation of nonlinear response. Figure 16 shows the dependence of nonlinear absorption (β) on the concentration of DBA and p-(N,N-dimethyl)DBA in solution. The NLA absorption coefficient β varies almost linearly with sample concentration. Similar behavior was observed with all other compounds.

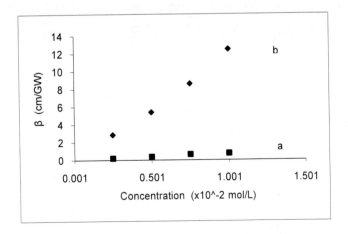

Figure 16. TPA coefficient β versus concentration of samples in DMF. a: DBA; b: p-(N,N-dimethyl)DBA.

The nonlinear absorption as well as the nonlinear refraction decreased as the concentration in solution decreased from 1×10^{-2} mol/L to 2.5×10^{-3} mol/L. It is also evident from figure 15 that optical limiting is better in PMMA doped with higher concentration of p-(N,N-dimethyl)DBA. Therefore, it is clear that the observed nonlinear response has direct relation with the concentration of samples in DMF.

Degenerate Four Wave Mixing

DFWM experiment was performed on p-(N,N-dimethyl)DBA using Nd: YAG 7 ns laser, in order to extract $\chi^{(3)}$ value and compare with that obtained through Z-scan technique. Variation of the DFWM signal as a function of the pump intensity for the compound dissolved in DMF at a concentration of 1×10^{-2} mol/L is shown in figure 17.

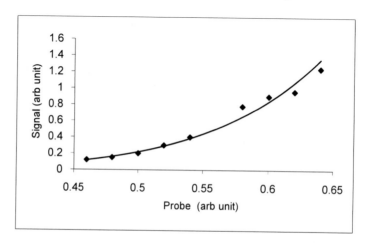

Figure 17. Phase conjugate signal versus probe for p-(N,N-dimethyl)DBA in DMF.

The signal is proportional to the cubic power of the input intensity as given by the equation

$$I(\omega)\alpha \left(\frac{\omega}{2\varepsilon_0 cn^2} \right)^2 \left| \chi^{(3)} \right|^2 l^2 I_0^3(\omega) , \qquad (7)$$

where $I(\omega)$ is the DFWM signal intensity, $I_0(\omega)$ is the pump intensity, l is the optical pathlength, and n is the refractive index of the medium. The solid curve in the figure is cubic fit to the experimental data. $\chi^{(3)}$ can be calculated from the equation:

$$\chi^{(3)} = \chi^{(3)}_{ref} \left[\frac{I/I_0^3}{(I/I_0^3)_{ref}} \right]^{1/2} \left[\frac{n}{n_{ref}} \right]^2 \frac{l_{ref}}{l} \left(\frac{\alpha l}{(1-e^{-\alpha l})e^{-\frac{\alpha l}{2}}} \right) \qquad (8)$$

where the subscript 'ref' refers to the standard reference CS_2 under identical conditions, α is the linear absorption coefficient of the sample and $\chi^{(3)}_{ref}$ is taken to be 4.0×10^{-13} esu [59, 60]. The $\chi^{(3)}$ value of p-(N,N-dimethyl)DBA is found to be 1.66×10^{-12} esu and it matches well with the value 1.9×10^{-12} esu obtained for the pure compound in DMF solution by Z-scan technique.

A check on the consistency of data in the experiment can be made by assuring that the conjugate signal varies as the appropriate power of the pump power or energy. The data can be fitted by a least square technique to the appropriate power law as discussed above. The closer the fit to the theory, the more accurate will be the value of $\chi^{(3)}$ obtained. If the linear absorption becomes substantial ($\alpha L \sim 1$), then equation (8) will no longer be appropriate, because the material is then not Kerr-like, and $\chi^{(3)}$ could be then due to population redistribution. Then the data should reflect this by showing a departure from a cubic dependence on pump intensity (average power or energy).

Figure of merit F is a measure of the nonlinear response that can be achieved for a given absorption loss, and is useful in comparing nonlinear materials in regions of absorption [60]. The figure of merit F is calculated by taking α in to account. F is given by

$$F = \chi^{(3)} \Big/ \alpha \tag{9}$$

F value calculated for p-(N,N-dimethyl)DBA is 1.608×10^{-11} esu cm, which is two order of magnitude larger than that measured for some organic systems by Reji Philip et al. [59].

The present study is focused on the third order NLO properties of chalcones in their solution state. However, it is worthwhile to mention here that, there is always scope for testing these results in their bulk phase, either in crystalline form or in the thin film form. Chalcones, though have low melting point, can be grown into crystals and Z-scan measurements can be repeated on them. Laser damage threshold of these crystalline materials can then be measured. The DFWM technique was employed in this study in order to compare $\chi^{(3)}$ values of samples with those obtained through Z-scan technique. However, one can also do time resolved measurements using DFWM technique.

CONCLUSION

We are introducing some chalcones as promising new third order nonlinear optical materials. Their third order nonlinear optical coefficients have been extracted through single beam Z-scan technique. An attempt has been made to explore and explain the relationships that exist between their molecular structures and the third order NLO properties. The third order nonlinear susceptibility as well as the second order hyperpolarizability of these molecules was found to vary with the extent of π-electron delocalization among them. All these molecules, except p-(N,N-dimethyl)DBA, exhibit strong two-photon absorption at 532 nm. p-(N,N-dimethyl)DBA exhibits strong reverse saturation absorption. A good optical limiting of nanosecond pulses has been demonstrated with these molecules. p-(N,N-dimethyl)DBA possesses a very high $\chi^{(3)}$ of the order of 10^{-12} esu that is comparable with those values obtained for tricyanovinyl azo dye and 4-(N, N-diethylamino)-4'-nitrostilbene (DEANS). A good comparison is observed between $\chi^{(3)}$ values of p-(N,N-dimethyl)DBA obtained through Z-scan and DFWM techniques. Present study shows that these chalcones show interesting third order response and are attractive for optical limiting and other photonic applications.

ACKNOWLEDGMENT

Authors wish to thank Dr. Reji Philip, Scientist, Optics Group, Raman Research Institute Bangalore, for providing laboratory facility to set up Four Wave Mixing experiment and for his fruitful suggestions. They also thank Mr. Suchand Sandeep, Research Scholar, Raman Research Institute, Bangalore, for his help in the DFWM experiments. Thanks to Prof. G. Umesh, Department of Physics, NITK Surathkal, for providing excellent laboratory facility for Z-scan experiment. The generosity of Prof. Shivarama Holla, Satheesh Rai N. and Mr. Mithun Ashok, Mangalore University with interesting chalcones has been gratefully acknowledged.

REFERENCES

[1]   Robert W. Boyd, *Nonlinear Optics*, Academic Press, Inc., New York, 1992.
[2]   Paras N. Prasad and David J. Williams *Introduction to Nonlinear Optical Effects in Molecules and Polymers,* John Wiley and Sons, New York, 1992.
[3]   J.A. Giordmaine, *Phys. Rev. Lett.* 1962, 8, 19.
[4]   P.D. Maker, R.W. Terhune, M. Nisenhoff, and C.M. Savage, *Phys. Rev. Lett.* 1962, 8, 21.
[5]   K. Uchida, S. Kaneko, S.Omi, C. Hata, H. Tanzi, Y. Asahara, A. J. Ikushima, T. Tokizaki, and A. Nakumara, *J.Opt.Soc. Am. B* 1994,11, 1236.
[6]   S. Matsumoto, K. Kubodera, T. Kurihara, and T. Kaino, *Opt. Commun.* 1990, 76, 147.
[7]   D. Ahn and S.L. Chuang, *IEEE J. Quant. Electron.* 1987, 23, 2196.
[8]   Seth R. Marder, William E. Torruellas, Mirelle Blanchard-Desce, Vincent Ricci, George I. Stegeman, Sharon Gilmour, Jean-Luc Bredas, Jun Li, Greg U. Bublitz, Steve G. Boxer, *Science* 1997, 276,1233.
[9]   J.W. Perry, K. Mansour, I. -Y. S. Lee, X. –L. Wu, P. V. Bedworth, C. –T. Chen, D.Ng, S.R.Marder,P. Miles, T.Wada, M. Tian, H. Sasabe, *Science* 1996, 273, 1533.
[10] Lee W. Tutt and Thomas F. Boggess *Prog. Quant. Electr.* 1993, 17, 299.
[11] L.A. Hornak, *Polymers for Light Wave and Integrated Optics Technology and Applications*, Chapter 21, Marcel Dekker Inc., New York, 1992.
[12] Q. Gong, Z. Xia, Y.H. Zou, X. Meng, L. Wei, and F.M. Li, *Appl. Phys. Lett.* 1991, 59, 381.
[13] J.E. Ehrlich, X.L. Wu, I.-Y.S. Lee, Z.-Y. Hu, H. Rockel, S.R. Marder, and J.W. Perry, *Opt. Lett.* 1997, 22, 1843.
[14] B. Sahraoui, X.N. Phu, M. Salle, and A. Gorgues, *Opt. Lett.* 1998, 23, 1811.

[15] Z. Yang, Z. Wu, J. Ma, A. Xia, Q. Li, C. Liu, and Q. Gong, *Appl. Phys. Lett.* 2005, 86, 061903.

[16] T. Kurihara, H. Kobayashi, K. Kubodera, and T. kaino, *Chem. Phys. Lett.* 1990, 165, 170.

[17] T. Kobayashi, H. Uchiki, and K. Minoshima, *Synth. Met.* 1989, 28, D699.

[18] T. Kurihara, H. Kobayashi, K. Kubodera, and T. kaino, *Opt. Commun.* 1991, 84, 149.

[19] H. Kanbara, H. Kobayashi, and K. Kubodera, *IEEE Photonics Technol. Lett.* 1989, PTL- 1, 149.

[20] T. Kurihara, Y. Mori, N. Ooba, S. Tomaru, and T. kaino, *J. Appl. Phys.* 1991, 70, 17.

[21] D.S. Chemla, J. Zyss (Eds), *Nonlinear Optical Properties of Organic Molecules and Crystals*, Vol 1, Academic Press, London, 1987.

[22] V. Crasta, V. Ravindrachari, R.F. Bhajantri, R. Gonsalves, *J. Crystal Growth*, 2004, 267, 129.

[23] V. Crasta, V. Ravindrachari, S. Lakshmi, S.N. Pramod, M.A. Shridar, and J.S. Prasad, *J. Crystal Growth* 2005, 275, e329.

[24] Di Wu, Bo Zhao, and Z. Zhou, *J. Mol. Stru. (Theochem)* 2004, 682, 83.

[25] G.J. Zhang, T. Kinoshita, K. Sasaki, Y. Goto, and M. Nakayama, *Appl. Phys. Lett.* 1990, 57, 221.

[26] B. S. Holla, B. Veerendra, J. Indira, *J. Crystal Growth 2003,* 252, 308.

[27] J. Indira, P.P. Karath, and B.K. Sarojini, *J. Crystal Growth* 2002, 242, 209.

[28] V. Ravindrachari, V. Crasta, R.F. Bhajantri, B. Poojari, *J. Crystal Growth* 2005, 275, e313.

[29] Y. Kitaoka, T. Sasaki, S. Nakai, A. Yokotani, Y. Goto, and M. Nakayama, *Appl. Phys. Lett.* 1990, 56, 2047.

[30] H. Kauffmann and F. Kieser, *Chem. Ber.* 1913, 46, 3788.

[31] W. Dilthey, L. Neuhaus, E. Reis, and W. Schommer, *J. Pract. Chem.* 1930, 124, 61.

[32] P. Pfeiffer and H. Kleu, *Chem. Ber.* 1933, 66B, 1058.

[33] G. Reddelein *Chem. Ber.* 1912, 45, 2904.

[34] V.G. Mitina, A.A. Sukhorukov, B.A. Zadorozhnyi and V. F. Lavrushin, Zn. *Obsch. Khim.* 1976, 46, 699.

[35] S. Ambelcar, S.S. Vernelcar, S.S. Acharya and S. Rajagopal, *J. Pharm. Pharmacol.* 1961, 13, 698.

[36] G. Wurm, Arch. Phar. 308 (1975) 142; R. Aries, German Patent 2,341,514 (1974), *Chem. Abstr.* 1974, 80, 1461522.

[37] N. P. Buu-Hoi, N.D. Xuong and M. Sy, *Bul. Soc. Chim. France* 1956, 1646.

[38] Benjamin, E. Z. Faraggi, Y. Anvy, D. Davidov and R. Neumann *Chem.Mater.* 1996, 8, 352.

[39] T. Cassano, R. Tomasi, M. Farrarra, F. Babudri, G. M. Farinola and F. naso *Chem. Phys.* 2001, 272, 111.

[40] R Schroeder, B Ulrich, W Graupner, and U Scherf *J. Phys.: Condens. Matter* 2001, 13, L313.

[41] Marius. Albota et. al., *Science* 1998, 281, 1653.

[42] John Kiran, K. Chandrasekharan, Satheesh Rai Nooji, H.D. Shashikala, G. Umesh and Balakrishna Kalluraya *Chem. Phys.* 2006, 324, 699.

[43] John Kiran A., Mithun Ashok, K. Chandrasekharan, B. S. Holla H. D. Shashikala and G. Umesh *Opt. Commun.* 2007, 269, 235.

[44]   M. Sheik-Bahae, Ali A. Said, Tai-Huei Wei, David J. Hagan, E. W. Van Stryland, *IEEE J. Quant. Electr.* 1990, 26, 760.

[45]   Richard L. Sutherland, *Handbook of Nonlinear Optics*, Marcel Dekker, Inc., New York 1996.

[46]   L. Zhi-Bo, T. J. Guo, Z. W. Ping, Z.W. Yuan, Z.G. Yin, *Chin. Phys. Lett.* 2003, 20, 509.

[47]   X. Liu, S. Guo, H. Wang, and L. Hou, *Opt. Commun.* 2001, 197, 431.

[48]   Y.R. Shen, *The principles of Nonlinear Optics,* Wiley-Interscience, New York, 1984.

[49]   T. Cassano, R. Tomasi, F. Babudri, M. Ferrarra, A. Cardone, G.M. Farinola and F. Naso, *Opt. Lett.* 2002, 27, 2176.

[50]   P.S. Kalsi, *Stereochemistry-Conformation and Mechanism*, Wiley Eastern Ltd, 1990.

[51]   S. –L. Guo, L. Xu, H. –T. Wang, X. –Z.You, and N. B. Ming, *Optik* 2003,114, 58.

[52]   G. S. He, G. C. Xu, P. N Prasad, B. A. Reinhardt, J. C. Bhatt, and A.G. Dillard, *Opt. Lett.* 1995, 20, 435.

[53]   T.F. Boggess Jr, K.M. Bohnert, K. Mansour, S.C. Moss, I.W. Boyd, and A.L. Smirl, *IEEE J. Quant. Electron.* 1986, QE-22, 360.

[54]   F. Z. Henari, W. J. Blau, L. R. Milgrom, G. Yahioglu, D. Philips and J.A. Lacey, *Chem. Phys. Lett.* 1997, 267, 229.

[55]   S. Couris, E. Koudoumas, A.A. Ruth, and S. Leach, *J. Phys. B: At. Mol. Opt. Phys.* 1995, 28, 4537.

[56]   M. Samoc, A. Samoc, B. Luther-Davis, *J Opt Soc Am B* 1998, 15, 817.

[57]   Billmeyer F.W.Jr., *Text Book of Polymer Science*, 3rd ed., John Willey and Sons, Singapore, 1994.

[58]   P.Yang, Xu J., J. Ballato, R.W. Schwartz, D.L. Carroll., *Appl Phy Lett* 2002, 80, 3394.

[59]   Reji Philip, M. Ravikanth, G. Ravindrakumar, *Opt. Commun.* 1999, 165, 91.

[60]   J.S. Shrik, J.R. Lindle, F.J. Bartoli, M.E. Boyle, *J Phys. Chem.* 1992, 96, 5847.

In: Perspectives in Optics Research
Editor: Jeffrey M. Ringer pp. 193-212

ISBN: 978-1-61122-934-9
© 2011 Nova Science Publishers, Inc.

Chapter 7

OPTICAL SUPERLATTICES: WHERE PHOTONS BEHAVE LIKE ELECTRONS

M. Ghulinyan, *Z. Gaburro, L. Pavesi*
Dipartimento di Fisica, University of Trento, Povo (Trento), Italy
C. J. Oton and N. Capuj
Departamento de Fisica Basica, University of La Laguna, Tenerife, Spain
R. Sapienza, C. Toninelli, P. Costantino, and D. S. Wiersma[†]
European Laboratory for Nonlinear Spectroscopy
and INFM-MATIS, Sesto Fiorentino (Florence), Italy

Abstract

We report on optical analogues of well-known electronic phenomena such as Bloch oscillations and electrical Zener breakdown. We describe and detail the experimental observation of Bloch oscillations and resonant Zener tunneling of light waves in static and time-resolved transmission measurements performed on optical superlattices. Optical superlattices are formed by one-dimensional photonic structures (coupled microcavities) of high optical quality and are specifically designed to represent a tilted photonic crystal band. In the tilted bands condition, the miniband of degenerate cavity modes turns into an optical Wannier-Stark ladder (WSL). This allows an ultrashort light pulse to bounce between the tilted photonic band edges and hence to perform Bloch oscillations, the period of which is defined by the frequency separation of the WSL states. When the superlattice is designed such that two minibands are formed within the stop band, at a critical value of the tilt of photonic bands the two WSLs couple within the superlattice structure. This results in a formation of a resonant tunneling channel in the minigap region, where the light transmission boosts from 0.3% to over 43%. The latter case describes the resonant Zener tunneling of light waves.

Keywords: optical superlattice, Wannier-Stark ladder, Bloch oscillations, Zener tunneling, porous silicon

[*]E-mail address: mghool@science.unitn.it; http://www.science.unitn.it/~semicon/
[†]E-mail address: wiersma@lens.unifi.it; http://www.complexphotonics.org/

1 Introduction

Transport phenomena of charge carriers in solids have been studied intensively. The energy spectrum of a quantum particle in a semiconductor crystal is described by extended Bloch modes and consists of allowed and forbidden bands. In the momentum space an electron moves in the allowed energy band running its quasi-impulse in the Brillouin zone. With a given energy, in the real space it travels freely through the infinite crystal.

When an external bias is applied on the crystal, the electron which travels from the Brillouin zone center towards the zone edge, is firstly accelerated by the field F. Then, upon reaching the Brillouin zone edge, its velocity continuously reduces down to zero and finally it starts to move in the opposite direction. Thus, in the Brillouin zone the electron performs a periodic motion, which can be described as an interference phenomenon brought up by Bragg reflection from the energy band edge once the particle wavelength reaches the crystal lattice period d. This periodic motion in k-space is accompanied by an oscillatory motion in real space, known as electronic Bloch oscillations [1], and has a characteristic period of $T_B = h/eFd$. The energy spectrum of such a particle in a semiconductor crystal, exposed to an external filed, is described by the so-called Wannier-Stark ladder (WSL) [2]. The WSL consists of a set of equidistant states, which are localized in space and have an energy separation proportional to the inverse of T_B.

The Bloch oscillations are closely related to another phenomenon, known as electrical breakdown or *nonresonant* Zener tunneling [3], which becomes effective as the electric field increases. In real space the tilt of the energy bands with increasing field gets stronger, and at high enough electric fields an electron tunnels to the continuum of states of an upper energy band without gaining extra energy from the field (this phenomenon has a wide application in microelectronic devices, such as the Zener and tunnel diodes). With the increasing escape probability of the electron to the other energy band the Bloch oscillations are damped strongly. On the other hand, *resonant* Zener tunneling is possible at high electric fields between charge carriers in WSLs.

The experimental study of electronic Bloch oscillations has been an issue for a long time because of the fact that in a semiconductor crystal an electron wave packet looses its coherence on a time scale, which is much shorter than the oscillation period T_B. The invention of electronic superlattices [4] opened up new possibilities to study such interference phenomena, which show dynamics faster than the decoherence time of an electron wave packet. This possibility is brought up by the large translational period of a superlattice structure, which implies a smaller Brillouin zone and, therefore, a shorter oscillation period. Recent experiments cover a series of results such as the observation of WSLs [5], Zener breakdown [6], and Bloch oscillations [7]. Resonant tunneling between the anticrossing Wannier-Stark states of neighboring energy minibands has been considered theoretically [8] and observed in experiments [9].

Various analogies between the transport of electrons and the propagation of light waves in dielectric materials have been established [10]. Electronic crystals have an analogue in the form of *photonic crystals*. Photonic crystals are artificial materials, in which a periodic variation of dielectric constant on a length scale comparable with the wavelength leads to the formation of bands where the propagation of photons is allowed or forbidden [11]. Since photons are uncharged particles and interact extremely weakly with each other, a light wave packet, that propagates through such a system, remains coherent for much longer times than charged particles. This means that dynamic interference effects could be isolated and studied more easily with light than with electrons.

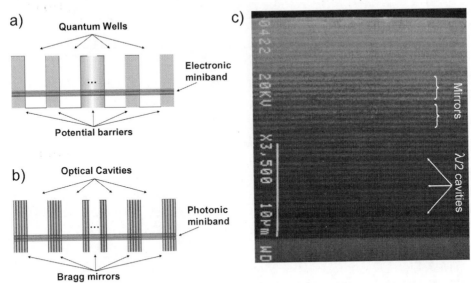

Figure 1: In analogy to the electronic coupling of separate quantum wells in a semiconductor superlattice (a), an optical superlattice can be realized when optical cavities are brought together (b) resulting in the formation of a miniband of extended photonic states. (c) A SEM micrograph of a one-dimensional porous silicon optical superlattice.

The existence of the optical counterpart of a WSL has been discussed theoretically [12] and observed experimentally in linearly chirped Moiré gratings [13]. Different photonic systems have been proposed to observe Bloch oscillations of light waves [14, 15]. Optical superlattices of coupled degenerate cavities have been proposed as a potentially ideal system to observe Bloch oscillations for light (see Fig. 1.1) [16]. In these structures an optical path gradient, $\Delta\delta$, *parallel* to the light propagation direction was suggested to mimic the optical equivalent of an external force (the static electric field in the electron case).

In this chapter we report on the observations of optical analogues of electronic Bloch oscillations and electrical Zener breakdown. We will describe and detail the experimental observation of Bloch oscillations [17] and resonant Zener tunneling [18] of light waves both in static and time-resolved transmission measurements performed on optical superlattices. The one-dimensional multilayer structures were made of porous silicon and were specifically designed to present a tilted photonic crystal band in close analogy to the tilted electronic miniband of a biased semiconductor superlattice. A controlled optical path gradient along the growth direction of the structure was used to form an optical Wannier-Stark ladder of equidistant photonic states. The frequency separation of these localized states defines the period of the photon Bloch oscillations. When the superlattice is designed such that two minibands are formed within the stop band, at a critical optical path gradient, we observe resonant coupling between two Wannier-Stark ladders, which leads to delocalization of the optical waves and, hence, resonant Zener tunneling.

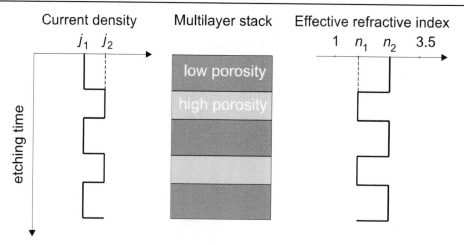

Figure 2: Multilayer growth in porous silicon: a modulation of the electrochemical current density in time (left) results in a formation of a stack of differently porous layers (middle), where each layer is described by an effective refractive index n_{eff} (right).

2   Porous Silicon-Based One-Dimensional Systems

The unique optical properties of porous silicon (PS) have attracted the researchers intensively in the last decade [19]. The electrochemical anodization procedure of silicon offers a cheap and fast technology for growing porous structures. Despite the structural inhomogeneities at the nanoscale, PS grown on heavily doped p-type silicon substrate shows straight columnar structures and passive optical properties of a dielectric material [20]. The typical pore sizes of 30-50 nm are more than an order of magnitude smaller than the wavelength of visible light, therefore an effective medium approximation allows one to characterize the porous structure with a single effective refractive index [1] n_{eff}, which is lower than the refractive index of bulk silicon.

    The porosity of a layer can be modulated by changing the current density during the etching process, which results in a modulation of n_{eff} (Fig. 1.2). An important property of the electrochemical etching is that silicon dissolution preferably takes place at the pore tips, where the local electric field has the highest value. This means that the already grown layers will not be affected by the changes in the current density, i.e. the pore sizes and the thickness of a layer will not change while new layers are growing. Then it becomes possible to realize one-dimensional multilayer structures, like dielectric Bragg mirrors and Fabry-Perót filters, by alternating high and low porosity layers. Usually, the preparation of PS structures of good optical quality, composed by several dozens of layers is a difficult task, because the anodization conditions might drift as the sample thickness increases [21]. These drifts affect the optical parameters of the layers, and one has to consider them when trying to maintain constant optical path in photonic structures. The reflectance spectrum is therefore heavily modified due to the inhomogeneities of the layers. Unfortunately, most

---

[1]There exists, however, some residual scattering especially at the interface between layers of different pore size

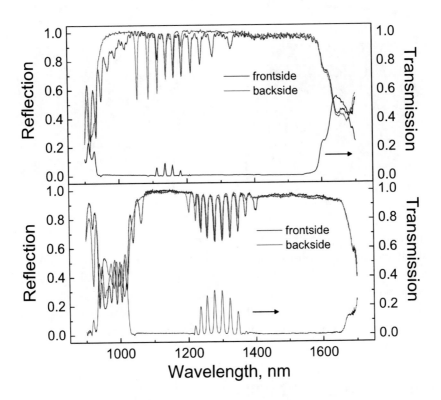

Figure 3: Reflectance (front- and backside) and transmittance spectra of a free-standing 10 coupled microcavities sample centered at 1300 nm: (top layer) naturally grown sample; (bottom layer) the same structure with compensated drift in layer thicknesses. The front- and backside reflectance spectra match in the compensated case, and the sample transmittance increases correspondingly, showing almost all features.

porous samples are not detached from the silicon substrate, limiting the investigation of the spectral properties of such photonic structures.

Free-standing PS-based multilayers [21], where both transmission and reflection measurements are possible, give richer information about the optical losses due not only to absorption but also to scattering, and allow to characterize the drifts in layer parameters (Fig. 1.3a). The natural drifts of the layer thickness and porosity can be successfully compensated by changing the etching parameters in a controlled way[2] (Fig. 1.3b). Moreover, as we will see in the following sections, we can use the optical path gradients for specific purposes and even enhance them in a controlled manner upon will.

---

[2]The optical thickness depends on both porosity and physical thickness of the layer. The physical thickness can be easily controlled by simply varying the duration of the etch, while the refractive index drift can be enhanced by varying the duration of so-called etch-stops (zero current pulses) combined with the use of a magnetic stirrer. Assuming a linear variation of the layer thickness with depth, one can compensate the evaluated drift by modifying the etching time so that a constant layer optical thickness with depth is attained.

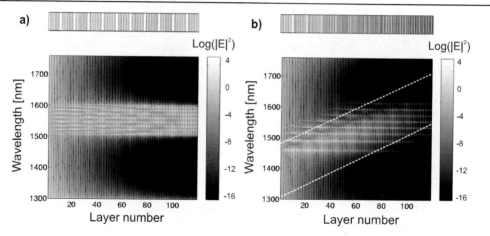

Figure 4: Light intensity distribution inside the optical superlattice structure. The parameters used in the calculations correspond to samples used in the actual experiment. (a) Flat band situation, $\Delta\delta = 0\%$, (b) tilted band situation, $\Delta\delta = 14\%$. The dashed lines are a guide to the eye that indicates the theoretical tilting of the miniband. Above each panel the coupled microcavity structure is schematically shown; the grey scale refers to the refractive index variation along the depth in the sample (the darker the larger n).

3 Bloch Oscillations of Light

3.1 The Optical Superlattice

In a close analogy to the formation of an electronic superlattice when several semiconductor quantum wells are brought together (Fig. 1.1(a)), an optical superlattice is made by coupling degenerate optical resonators (cavities) within the same photonic structure (Fig. 1.1(b))[21, 22]. Optical coupling between the various cavities yields a degenerate mode repulsion [23] and a formation of a miniband of optical states, which are densely packed around the resonant wavelength. In one dimension, an optical superlattice can be formed by stacking two dielectric layers A and B with different refractive indices and quarter-wave thickness in a way to form identical cavities separated by Bragg mirrors (Fig. 1.1(c)). In particular, we choose the following sequence of layers: $BABABABAB \, (AA)_1$ $BABABABAB \, (AA)_2 \ldots (AA)_m \, BABABABAB$. This structure is essentially a series of m microcavities $(AA)_m$ coupled to each other through the $BABABABAB$ Bragg mirrors. The amount of splitting of cavity resonances is given by the strength of the coupling mirrors[3], i.e. the weaker the mirrors are, the larger splitting occurs.

The light intensity distribution inside this structure can be calculated using a transfer matrix (TM) formalism [24]. Figure 1.4(a) shows the appearance of a miniband of extended optical states (bright lines) in the photonic stop band (dark regions) in the conditions of a constant optical path throughout the structure.

---

[3]For a given index contrast between A and B layers, the reflectance of a Bragg mirror improves with increasing the number of periods. Similarly, for a fixed number of periods, the mirror reflectance grows with an increase in the refractive index contrast

In order to obtain optical Bloch oscillations one has to introduce a gradient in the optical thickness of the layers. This will result in a spatial tilting of the miniband (Fig. 1.4(b)) and in the formation of an optical Wannier-Stark ladder. The latter manifests as a series of narrow equidistant transmission peaks with a frequency separation of the peaks that defines the period T_B of the Bloch oscillations. The linear change of the optical thickness introduces, to first order, a linear tilt of the miniband [25]. The oscillation period will decrease with increase in the miniband tilting.

3.2   Sample Preparation and Optical WSLs

We have grown the optical superlattices by controlled electrochemical etching of heavily doped p-type (100)-oriented silicon. The electrolyte was prepared mixing a 30 % volumetric fraction of aqueous HF (48 wt.%) with ethanol. A magnetic stirrer was used to improve electrolyte exchange. The applied current density defined the porosity of the layer. We applied 50 mA/cm^2 for the high porosity layer A (porosity 73%, refractive index $n_A = 1.45$) and 7 mA/cm^2 for the low porosity layer B (50%, $n_B = 2.1$). The physical thickness of the layers was controlled by adjusting the duration of the etch times. Alternating these currents, we grew a superlattice structure with ten coupled cavities. The superlattice structures were made free-standing by applying an electropolishing current pulse at the end of the growth process [21]. Particular care was taken to control the anodization conditions which usually drift as the total sample thickness increases. Moreover, the natural refractive index drift was compensated by changing the etching times of the layers. The process is known to provide excellent control over the layer properties allowing the growth of Fabry-Perót filters with a resonance quality factors up to 3300 [22].

The one dimensional translational symmetry of the superlattice was broken by introducing an optical path gradient in the growth direction of the structure. This was achieved by changing the duration of the etch stop current, which controlled the refractive index and hence the variation in the optical thickness of each layer. We produced samples with gradient values in the range from $\Delta\delta = 2$ to 14% (values that were extracted from the best fit parameters of the transmission spectra). As porous silicon samples suffer from lateral inhomogeneities due to doping variations of the silicon wafers, an inhomogeneous widening of the transmission peaks is usually measured when broad probe beams are used. In order to avoid this, some spectra were measured in a high-resolution transmission setup with a very small numerical aperture (NA ~ 0.0075, leading to a negligible broadening of 0.02 nm at 1550 nm wavelength), where a tunable laser source focused to a $35\mu m$ diameter spot was used.

Figure 1.5 shows the transmission spectra for different values of gradient $\Delta\delta$. The flat miniband situation is reported in the top panel (Fig. 1.5(a)). Optical WSLs are formed with an increase in the gradient over a certain value: the bigger is $\Delta\delta$, the larger is the energy separation of the Wannier-Stark states. The states in an optical Wannier-Stark ladder are not extended and have a reduced spatial extent inside the superlattice structure. The transmission resonances become narrower and less intense for the large gradient values (Fig. 1.5(b-d)). This is a clear signature of the strongly inclined miniband, which also means that in order to be transmitted the photons now need to overcome a thicker tunneling barrier.

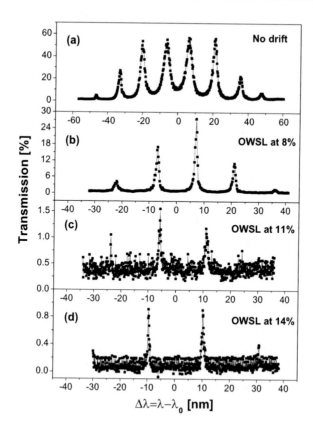

Figure 5: High resolution transmission spectra of the optical superlattices with different gradients of the optical thickness of the layers ($\lambda_0 = 1.55\mu m$ is the central wavelength). The top spectrum corresponds to the non drifted sample (spatially flat miniband), while the others (b), (c), (d) show the occurrence of the optical Wannier-Stark ladder with equidistant resonances: the energy separation of the states increases with the increase of the miniband tilt.

3.3 Time-Resolved Transmission Measurements

We have performed time-resolved transmission experiments on these samples using an optical gating technique. This technique involves mixing a reference beam together with the transmitted signal in a 0.3mm thick non-linear crystal (β-Barium Borate) to produce a sum frequency signal. The probe beam is obtained from an optical parametric oscillator (Spectraphysics OPAL) pumped by a Ti:sapphire laser at center wavelength 810 nm (pulse duration 130 fs, average power 2.0 W, repetition rate 82 MHz) yielding short pulses tunable from 1300 to 1600 nm wavelength range (average power 100 mW). The reference pulse at 810 nm wavelength is obtained from the residual Ti:sapphire beam (450 mW average power). The sum frequency signal is detected by a photodiode, and a standard lock-in technique is used to suppress noise. A delay line in the reference beam allows to tune the time delay between signal and reference and thus to scan the signal pulse in time. In the top panel

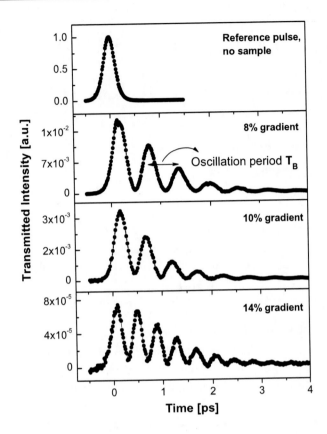

Figure 6: Time-resolved transmitted signals through the superlattice structures for various values of the optical path gradient $\Delta\delta$. The top panel shows the undisturbed probe pulse without sample. An oscillating signal is measured in the photodetector when the optical WSL are excited. The period of the oscillations and the total transmission decrease while increasing $\Delta\delta$.

of Fig. 1.6 we plot the system response without sample from which we determine that the temporal resolution of our system is smaller than 250 fs. The apparatus is designed such that the transmission spectrum of the sample can be monitored during the time-resolved measurement by sampling the transmitted light.

Time-resolved data indeed show that transmitted signals oscillate with a period T_B that decreases as the band tilting increases (Fig. 1.6). It is important to note that the optical WSL in our structures is formed above a threshold gradient value ($\sim7\%$). This is confirmed also by the TM-calculations. The reason is that an optical thickness gradient value below 7% is not sufficient to tilt fully the miniband within our sample thickness. One can see how the amplitude of the transmitted intensity decreases as the gradient increases: as a result of an increased tilt of the photonic miniband (see also Fig. 1.4(b)) the oscillating photon wave packet has to tunnel out through a thicker barrier. In Fig. 1.7 the measured periods of the Bloch oscillations are compared to the ones calculated through TM formalism. The

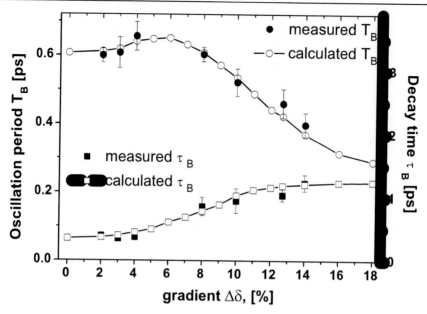

Figure 7: Measured photonic Bloch oscillation period T_B and decay time τ_B as a function of the gradient $\Delta\delta$. The error bars are the standard deviations obtained from various measurements on several positions on the sample and represent therefore the effect of lateral sample inhomogeneities. The solid lines are calculated using the transfer matrix method.

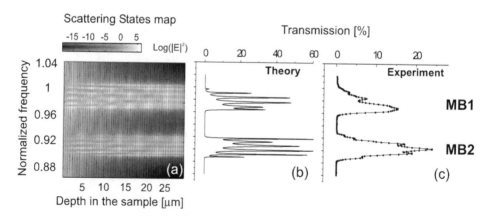

Figure 8: The calculated intensity distribution of the light inside the double miniband sample plotted as a color scale versus the normalized frequency ω/ω_0, where ω_0 is the minigap central frequency, and the depth inside the sample. (a) Flat miniband situation, $\Delta\delta = 0\%$. Two minibands MB1 and MB2, separated by a minigap region are seen in the calculated (b) and measured transmission spectrum (c).

experimental data are in perfect agreement with the theoretical prediction. However, one can measure oscillations even below the critical gradient value of 7%, which in this case are due to the reflection of light from the sample physical boundaries, therefore changing the gradient in this range does not influence the oscillation period. At larger values of gradient the period of Bloch oscillations decreases linearly with the increase in the miniband tilt, as expected.

The oscillations decay with a characteristic time τ_B which saturates at large gradients. This is a sign of the increased confinement of the optical modes in the WSL: as the photonic band tilt gets steeper, tunneling out of the sample becomes more difficult, and the transmission losses decrease accordingly. These intrinsic losses are not the only one present in the structure, as light in porous silicon suffers also from external losses. At large gradient values τ_B saturates to 1.26 ps, which is caused by scattering on the pores and residual absorption losses in the porous silicon sample. One can consider that $\tau_B = (\tau_{pBO}^{-1} + \tau_{ext}^{-1})^{-1}$, where τ_{pBO} is the intrinsic decay time of the Bloch oscillation and τ_{ext} is due to absorption and scattering losses. The solid line in Fig. 1.7 is obtained by assuming τ_{ext} =1.3 ps that corresponds to an extinction coefficient (total losses, absorption + scattering) of α_{ext} =100 cm^{-1}, in agreement with previously determined loss values [21].

4   Resonant Zener Tunneling of Light

4.1   The Double-Miniband Optical Superlattice

The successful demonstration of the optical analogue of electronic Bloch oscillations naturally raises the question whether it is possible to observe the optical analogue of Zener breakdown phenomenon. As a first reasoning towards the study of Zener tunneling of light, one could think of tilting the optical superlattice bands such that the first order (centered at λ_c) miniband is coupled to the next order one (centered at $\lambda_c/3$). However, this would require very large band tilting, and the realization of such structures, where the optical parameters are controlled and characterized, is impossible in practice. A possible solution is to build an optical superlattice that exhibits two minibands within the stop band. This will allow to use relatively small optical path gradients. For this, one should couple within the same structure two sets of cavities of C and D type, which are centered at different wavelengths λ_1 and λ_2, respectively[4]. The appropriate sequence of layers looks like the following: $BABABAB\,(CC)_1\,BABABAB\,(DD)_1\,\ldots\,(CC)_m\,BABABAB\,(DD)_m\,BABABAB$. In our studies we have chosen m=6, $\lambda_1 = 0.81\lambda_c$ and $\lambda_2 = 0.88\lambda_c$.

The samples were grown using the same technique described in the previous section. The refractive indices for this type of structures were determined to be $n_A = n_C = n_D = 1.5$ and $n_B = 2.12$. We produced samples with controlled gradient values in the range from $\Delta\delta$ = 0 to 18%. The light intensity distribution inside this structure in the absence of optical path gradient (Fig. 1.8(a)) shows the formation of two flat minibands MB1 and MB2. These appear as two sets of intense lines of extended optical states, stretching through the sample. The two minibands are separated by a photonic minigap (dark region), which shows negligible transmission of 0.43%. The parameters used in the calculations correspond to those of the samples studied in the actual experiment. The calculated spectra also take into account a loss coefficient of 100 cm^{-1} (due to absorption and scattering out of the

---

[4]Note that C and D layers are also quaterwave-thick, so that (CC) and (DD) form $\lambda_1/2$ and $\lambda_2/2$ cavities, respectively.

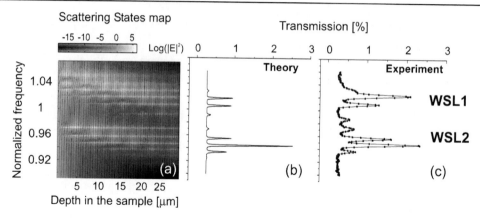

Figure 9: (a) Optical WSLs of localized modes are formed in two minibands at $\Delta\delta = 6.7\%$, which appear as weak narrow peaks in the transmission spectra (graphs in the right). Light transmission drops down to 2%: transfer matrix calculations (b) and the experimental data (c).

propagation axis) which gives a nearly negligible spectral broadening of the transmission peaks of $0.3 - 0.5$ nm. The corresponding calculated transmission spectrum (Fig. 1.8(b)) is compared with the experimental one (Fig. 1.8(c)). The latter is measured with a Varian Cary 5000 spectrophotometer using a broad beam of 1mm in diameter, which explains the appearance of wider and less intense spectral features. The high-resolution transmission setup was utilized in some particular cases, where the observation of narrow peaks with high intensity was essential.

An introduction of 6.7% of negative optical path gradient[26] tilts the photonic band structure and results in the formation of optical WSLs in both minibands (Fig. 1.9(a)). Now the spatial confinement of the localized states causes a decrease of the absolute transmission from 50% in the flat band case (delocalized states) down to 2% (Fig. 1.9(b)). In Fig. 1.9(c) we plot the measured transmission spectrum of the superlattice with tilted minibands, where the WSLs can be appreciated.

We go on further increasing the optical path gradient and we look at the evolution of photonic miniband picture. At a *critical* degree of band tilting ($\Delta\delta \sim 10.3\%$ in our case) the WSLs in two minibands couple within the extension of the structure. Coupling induced delocalization of two anticrossing states takes place, which appears as an intense resonant tunneling channel (Fig. 1.10(a)). The resonant Zener tunneling appears as an enhanced transmission peak in both calculated (Fig. 1.10(b)) and measured transmission spectra (Fig. 1.10(c)).

This observation, together with the TM-calculations in Figs. 1.8-1.10, nicely demonstrates the physics of resonant Zener tunneling. An efficient transmission channel opens when two internal resonances couple to form a delocalized mode, which has a transmission coefficient much larger than the sum of the transmission coefficients of the two individual resonances (before coupling). Such internal resonances can only couple if the frequency difference between the modes is smaller than their bandwidth. The characteristic property of Zener tunneling is that this frequency difference is tuned by changing the gradient inside

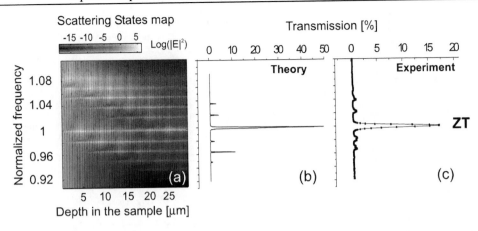

Figure 10: (a) The two WSLs couple at $\Delta\delta = 10.3\%$ and form a resonant tunneling channel through the sample. Resonant Zener tunneling is predicted by theory (b) and confirmed experimentally (c) as an enhanced transmission peak in the center of the minigap.

the sample and that the internal resonances arise from a double Wannier-Stark ladder. The critical gradient value $\Delta\delta_{ZT}$ at which Zener tunneling occurs (in our case 10.3%) depends on the frequency difference between the centers of the minigaps $\Delta\omega = \omega_1 - \omega_2$, and the bandwidth of the minibands themselves. At $\Delta\delta_{ZT}$ the frequency of the low frequency miniband at the last cavity (high depth) matches the frequency of the high frequency miniband at the first cavity (small depth). In Fig. 1.11 we show some numerical results which reflect the evolution of the optical WSLs towards the eventual coupling: the shrinkage of the minigap is plotted versus the increasing $\Delta\delta_{ZT}$. The calculations are performed for different reflectivities of Bragg mirrors (number of mirror periods) in the optical superlattice. In the limit of very small bandwidth (Bragg mirrors consisting of large number of layers resulting in weak coupling between the cavities) the optical thickness gradient at which Zener tunneling occurs takes its minimum value defined as $\Delta\delta_{ZT} = \Delta\omega/\omega_1$ (solid line). At larger bandwidths the situation is more complicated and to obtain $\Delta\delta_{ZT}$ one needs to calculate the scattering states map at each gradient value as in Fig. 1.4. We will see that the theoretical value for $\Delta\delta_{ZT}$ is indeed 10.3% for our sample.

We can observe from the graph in Fig. 1.11 that for the case of strong coupling between the cavities (1.5- and 2.5-period mirror cases) the relative shift of the resonance is negative for small gradients. This is due to the fact that the bandwidth of the miniband decreases slightly when the small gradient only detunes the coupling mirrors but is still not enough to break the translational symmetry of the superlattice. Over a certain value WSLs start to form and the minibands start to expand monotonically with the further increase of titling. Another interesting observation is the behavior of the curve for the strongest coupling case (1.5-periods) in the vicinity of $\Delta\delta_{ZT} \approx 18\%$: the coupling between the two WSL states is so strong that a change in the tilting of the miniband around $\Delta\delta_{ZT}$ does not influence the relative shift of the resonance. This means that a marked double peak would manifest in the transmission spectrum for all this range of gradients because of enhanced level repulsion between the degenerate states.

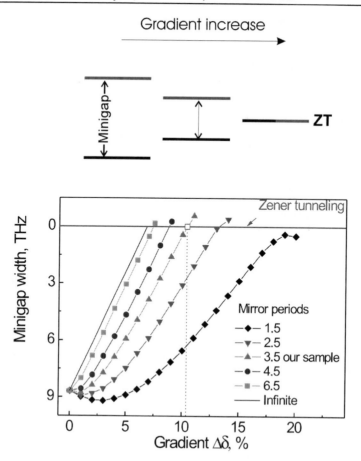

Figure 11: The calculated minigap width reduction is plotted versus the increasing gradient of the optical path for various reflectivities of coupling mirrors (number of periods). In the limit of very weak coupling between the cavities (solid line) the optical thickness gradient at which Zener tunneling occurs has a minimum value. The sketch in the top describes the minigap width reduction.

We decided to examine the Zener tunneling regime in detail. For this we have grown a superlattice structure with *lateral* (in-plain) variation in $\Delta\delta$ around 10.3%. To achieve this, the magnetic stirrer was placed close to one edge of the sample during the whole growth process, thus the electrolyte exchange was enhanced differently between the closest (to the stirrer) and farthest points on the sample surface. The latter resulted in a smoothly varying refractive index profile between the antipodal points in each layer of the optical superlattice (see the schematic sketch in Fig. 1.12). Measuring the transmission spectrum at different points over the sample surface (moving the incident beam laterally) allowed us to follow the evolution of miniband coupling around $\Delta\delta_{ZT}$. In this way the system was studied at different values of gradient between 6.5% to 10.7%.

High-resolution transmission measurements are performed for this sample. Fig. 1.12 reports the measured high-resolution transmission spectra taken at different points corre-

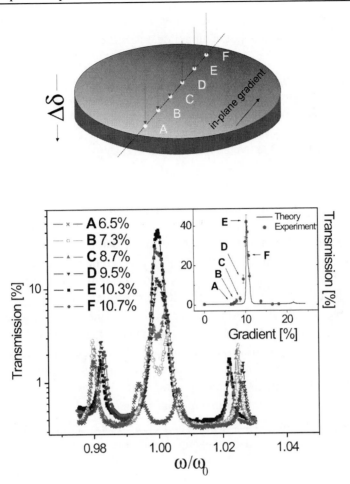

Figure 12: (top) A schematic view of the Hi-Res transmission measurement of the super-lattice sample with in-plane gradient. (bottom) The comparison between the experimental transmission values of the maximum transmission (line) around the central frequency ω_0 as a function of the gradient and the transfer matrix calculations (dots). The inset shows the transmission spectrum of the tilted superlattice around the value of the optical path gradient where the first anti-crossing of the optical Wannier-Stark ladders and hence Zener tunneling occurs.

sponding to different gradients. In the vicinity of the threshold gradient the transmission spectrum is very sensitive to small changes of the optical path. One can see how the edge states of the two WSLs start to overlap, and the saddle-like curvature transforms gradually into a sharp resonance of 42% transmission. In the inset of Fig. 1.12 the calculated transmission values at ω_0 for $\Delta\delta$ in the range $0 \div 25\%$ are compared with the experimental data. The correspondence between the experimental and calculated values of the Zener tunneling gradient is very good. In the calculations another maximum, corresponding to the second anticrossing of WSL states, is present at $\sim 21\%$. The intensity of this peak is much weaker compared with the first one at $\Delta\delta_{ZT} = 10.3\%$, which can be understood in terms of decreasing probability of the photon tunneling out of the superlattice as the miniband tilt increases.

The analysis of spectra around the Zener tunneling regime shows that, together with the increase of the transmission value, the resonant transmission peak gets wider[5]. At resonance its full width half maximum (FWHM) is roughly a factor of two larger than that of the uncoupled WSL peaks at $0.98\omega_0$ and $1.025\omega_0$. This broader lineshape is the effect of repulsion between the two coupled resonances.

4.2   Resonant Zener Tunneling in Time-Domain

When studying narrow spectral features in static transmission measurements, one usually is limited by the setup resolution. Therefore, lineshape differences are often difficult to appreciate even in the high precision spectra. Small variations in the frequency domain result in big fluctuations in the time domain. The Fourier transformation of a single resonance, corresponding to a phase shift of π, results in an asymmetric lineshape with single exponential decay at long times. A double resonance mode lineshape experiences a different phase shift (2π), which affects the shape of the time response, in particular, shifting its maximum towards longer times. Also, the Fourier transformation of a double resonance peak results in a more symmetric pulseshape in the time domain. This feature should allow to distinguish among peaks of the same width, originating from resonances of different multiplicity.

We have tested these properties looking at the time response of our optical superlattice sample, performing ultrashort pulse propagation experiment in the Zener tunneling regime. The technique used is the same as that described in detail for the Bloch oscillations case. Figure 1.13 shows three examples of transmitted pulses centered at different wavelengths. A reference pulse, measured in the absence of sample is plotted as dotted line for comparison. When a single Wannier-Stark state is excited (Fig. 1.13(a)), the transmitted signal intensity decays exponentially, which is the characteristic behavior of a localized state. The delay of the pulse, defined as the delay of the center of the mass of the pulse profile, as expected, is not big. When the incident pulse excites two resonances, a complex signal oscillating with a period of ~ 300 fs, determined by the frequency separation of the excited states, is observed (Fig. 1.13(b)). These are the well known photonic Bloch oscillations. In our specific case, this the oscillations are damped because of the strong coupling of the double resonance with the environment. Finally, Fig. 1.13(c) shows the time response of the peak with enhanced transmission at 1560 nm. The time-resolved profile does not have the typical shape of a single resonance. The observed picture is consistent with a double resonance transport behavior: it is characterized by a rapid decay and a substantial delay of maximum transmission point that amounts to almost 360 fs. The fast decay time is due to the strong coupling of the mode with the sample environment. The delay is caused by the transient time necessary to build up the double resonance of the Zener tunneling mode.

---

[5]The resonance lifetime, which is inversely proportional to the FWHM, decreases for the coupled states; which provides an additional proof of coupling induced delocalization of WSL states in the Zener tunneling regime.

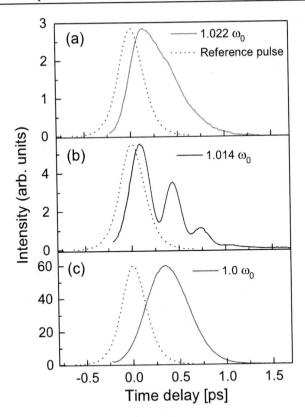

Figure 13: Time-resolved transmitted signals from a double miniband superlattice with optical path gradient in the Zener tunneling regime. The panels (a), (b), and (c) correspond to different probe wavelengths. (a) A excited single resonance shows a characteristic exponential decay whereas (b) damped Bloch oscillations are observed when exciting two Wannier-Stark resonances. In (c) the Zener tunneling peak is excited leading to a strongly delayed but nearly symmetric transmitted pulse. The dotted curves refer to the transmission in the absence of a sample.

5 Conclusion

In conclusion we have observed the optical counterparts of electronic Bloch oscillations and Zener tunneling in optical superlattices of porous silicon. A linear variation in the optical constants of the system along the propagation direction allows to form an optical Wannier-Stark ladder and to observe photonic Bloch oscillations resolved in time. Both the oscillation period and the damping time versus the strength of the Wannier-Stark ladder are consistent with predictions from transfer matrix calculations. The observation of resonant Zener tunneling of light waves has been performed via spectral and time-resolved transmission measurements on specifically designed optical superlattices, which exhibit two minibands. At a critical gradient of the optical path value, coupling of photonic minibands occurs resulting in delocalization of the optical Wannier-Stark states and, consequently, Zener tunneling of the light waves. The transition from low to high transmission is extremely sensitive to the variation of optical path gradient, making a Zener tunneling light valve a strong possibility. These fascinating parallels between electrons and photons not only show the potential of complex photonic structures to study fundamental problems, but also add new functionalities to silicon.

Acknowledgments

The authors would like to thank the financial support by the EC under contract number IST-2-511616 and MIUR through COFIN projects "Silicon-based photonic crystals: technology, optical properties and theory" and "Silicon-based photonic crystals for the control of light propagation and emission" and FIRB projects "Sistemi Miniaturizzati per Elettronica e Fotonica" and "Nanostrutture molecolari ibride organiche-inorganiche per fotonica".

References

[1] F. Bloch, Über die Quantenmechanik der Elektronen in Kristallgittern, *Z. Phys.* **52**, 555-6000 (1928).

[2] G.H. Wannier, Possibility of a Zener Effect, *Phys. Rev.* **100**, 1227 (1955).

[3] C. Zener, A theory of the electrical breakdown of solid dielectrics, *Proc. R. Soc. London Ser. A* **145**, 523-529 (1934).

[4] L. Esaki and R. Tsu, Superlattice and negative differential conductivity in semiconductors, *IBM J. Res. Dev.* **14**, 61-65 (1970).

[5] E.E. Mendez, F. Agullo-Rueda, and J.M. Hong, Stark localization in GaAs-GaAlAs superlattices under an electric field, *Phys. Rev. Lett.* **60**, 2426-2429 (1988).

[6] H. Schneider, H.T. Grahn, K.V. Klitzing, K. Ploog, Resonance-Induced Delocalization of Electrons in GaAs-AlAs Superlattices, *Phys. Rev. Lett.* **65**, 2720-2723 (1990); B. Rosam, D. Meinhold, F. Löser, V.G. Lyssenko, S. Glutsch, F. Bechstedt, F. Rossi, K. Köhler, and K. Leo, Field-Induced Delocalization and Zener Breakdown in Semiconductor Superlattices, *Phys. Rev. Lett.* **86**, 1307-1310 (2001).

[7] T. Dekorsy, P. Leisching, K. Köhler, and H. Kurz, Electro-optic detection of Bloch oscillations, *Phys. Rev. B* **50**, 8106-8109 (1994); V.G. Lyssenko, G. Valusis, F. Löser, T. Hasche and K. Leo, Direct Measurement of the Spatial Displacement of Bloch-Oscillating Electrons in Semiconductor Superlattices, *Phys. Rev. Lett.* **79**, 301-304 (1997); T. Hartmann, F. Keck, H. J. Korsch and S. Mossmann, Dynamics of Bloch oscillations, *New J. Phys.* **6**, 2-25 (2004).

[8] S. Glutsch and F. Bechstedt, Interaction of Wannier-Stark ladders and electrical breakdown in superlattices, *Phys. Rev. B* **60**, 16584-16590 (1999).

[9] B. Rosam, K. Leo, M. Glück, F. Keck, H. J. Korsch, F. Zimmer, K. Köhler, Lifetime of Wannier-Stark states in semiconductor superlattices under strong Zener tunneling to above-barrier bands, *Phys. Rev. B* **68**, 125301 (2003).

[10] Ping Sheng, *Introduction to Wave Scattering, Localization, and Mesoscopic Phenomena*, Academic Press, New York, (1995); *Wave Scattering in Complex Media, from theory to applications*, edited by S.E. Skipetrov and B.A. van Tiggelen, NATO series II, Vol. 107 (Kluwer, Dordrecht, 2003).

[11] J.D. Joannopoulos, R.D. Meade, and J.N. Winn, *Photonic Crystals* (Princeton University Press, Princeton, NJ, 1995); *Photonic Crystals and Light Localization in the 21st Century*, Nato Advanced Research Institute, series C, 563, edited by C.M. Soukoulis (Kluwer, Dordrecht, 2001).

[12] G. Monsivais, M. del Castillo-Mussot, and F. Claro, Stark-Ladder Resonances in the Propagation of Electromagnetic Waves, *Phys. Rev. Lett.* **64**, 1433-1436 (1990).

[13] C. Martijn de Sterke, J.N. Bright, P. A. Krug, and T. E. Hammon, Observation of an optical Wannier-Stark ladder, *Phys. Rev. E* **57**, 2365-2370 (1998).

[14] G. Lenz, I. Talanina, and C. Martijn de Sterke, Bloch Oscillations in an Array of Curved Optical Waveguides, *Phys. Rev. Lett.* **83** , 963-966 (1999); A. Kavokin, G. Malpuech, A. Di Carlo, P. Lugli, F. Rossi, Photonic Bloch oscillations in laterally confined Bragg mirrors, *Phys. Rev. B* **61**, 4413-4416 (2000).

[15] P.B. Wilkinson, Photonic Bloch oscillations and Wannier-Stark ladders in exponentially chirped Bragg gratings, *Phys. Rev. E* **65**, 56616 (2002).

[16] G. Malpuech, A. Kavokin, G. Panzarini, and A. Di Carlo, Theory of photon Bloch oscillations in photonic crystals, *Phys. Rev* **B**63 , 035108 (2001).

[17] R. Sapienza , P. Costantino, D.S. Wiersma, M. Ghulinyan, C.J. Oton, and L. Pavesi, Optical Analogue of Electronic Bloch Oscillations, *Phys. Rev. Lett.* **91**, 263902 (2003).

[18] M. Ghulinyan, C.J. Oton, Z. Gaburro, L. Pavesi, C. Toninelli, D.S. Wiersma, Zener Tunneling of LightWaves in an Optical Superlattice, *Phys. Rev. Lett.* **94**, 127401 (2005).

[19] S. Ossicini, L. Pavesi, F. Priolo, *Light Emitting Silicon for Microphotonics* , Springer Tracts in Modern Physics, Vol. 194 (Springer, Berlin, 2003); *Properties of Porous Silicon*, edited by L. Canham, (Short Run Press Ltd., London, 1997).

[20] C.J. Oton, M. Ghulinyan, Z. Gaburro, P. Bettotti, L. Pavesi, L. Pancheri, S. Gialanella, N.E. Capuj, Scattering rings as a tool for birefringence measurements in porous silicon, *J. Appl. Phys.* **94**, 6334-6340 (2003).

[21] M. Ghulinyan, C. J. Oton, Z. Gaburro, P. Bettotti, and L. Pavesi, Porous silicon freestanding coupled microcavities, *Appl. Phys. Lett.* **82**, 1550-1552 (2003).

[22] M. Ghulinyan, C. J. Oton, G. Bonetti, Z. Gaburro, and L. Pavesi, Free-standing porous silicon single and multiple optical cavities, *J. Appl. Phys.* **93**, 9724-9729 (2003).

[23] L. Pavesi, G. Panzarini, and L.C. Andreani, All-porous silicon-coupled microcavities: Experiment versus theory, *Phys. Rev. B* **58**, 15794-15800 (1998).

[24] J.B. Pendry, Symmetry and transport of waves in one-dimensional disordered systems, *Adv. Phys.* **43**, 461-542 (1994).

[25] Let us consider the flat bandedge energy, $E_0 \sim 1/\lambda_0$ and the tilted bandedge one, $E \sim 1/\lambda$, where $\lambda = \lambda_0(1+\Delta\delta)$, with $\Delta\delta$ the optical path gradient. Then for the ratio $\Delta E/E_0$ (where $\Delta E = E - E_0$) one can have $\Delta E/E_0 = 1/(1 + \Delta\delta) - 1$, which for small $\Delta\delta$ is a linear relation.

[26] For technical reasons, the gradient is defined with reference to the last layer. This causes a small shift of the spectral features to higher frequencies when the gradient is increased. A gradient of 6.7% therefore means that the first layer (depth zero) has an optical thickness that is 6.7% smaller than the optical thickness of the last layer.

INDEX

D